Algebra 2

LARSON
BOSWELL
KANOLD
STIFF

Applications • Equations • Graphs

Chapter 6
Resource Book

The Resource Book contains the wide variety
of blackline masters available for Chapter 6.
The blacklines are organized by lesson. Included
are support materials for the teacher as well as
practice, activities, applications, and assessment
resources.

McDougal Littell
A HOUGHTON MIFFLIN COMPANY
Evanston, Illinois • Boston • Dallas

Contributing Authors

The authors wish to thank the following individuals for their contributions to the Chapter 6 Resource Book.

Rose Elaine Carbone

José Castro

John Graham

Fr. Chris M. Hamlett

Edward H. Kuhar

Cheryl A. Leech

Ann C. Nagosky

Karen Ostaffe

Leslie Palmer

Ann Larson Quinn, Ph. D.

Chris Thibaudeau

ISBN: 0-618-02014-4

Contents

6 Polynomials and Polynomial Functions

Contents

Contents

Descriptions of Resources

This Chapter Resource Book is organized by lessons within the chapter in order to make your planning easier. The following materials are provided:

Tips for New Teachers These teaching notes provide both new and experienced teachers with useful teaching tips for each lesson, including tips about common errors and inclusion.

Parent Guide for Student Success This guide helps parents contribute to student success by providing an overview of the chapter along with questions and activities for parents and students to work on together.

Prerequisite Skills Review Worked-out examples are provided to review the prerequisite skills highlighted on the Study Guide page at the beginning of the chapter. Additional practice is included with each worked-out example.

Strategies for Reading Mathematics The first page teaches reading strategies to be applied to the current chapter and to later chapters. The second page is a visual glossary of key vocabulary.

Lesson Plans and Lesson Plans for Block Scheduling This planning template helps teachers select the materials they will use to teach each lesson from among the variety of materials available for the lesson. The block-scheduling version provides additional information about pacing.

Warm-Up Exercises and Daily Homework Quiz The warm-ups cover prerequisite skills that help prepare students for a given lesson. The quiz assesses students on the content of the previous lesson. (Transparencies also available)

Activity Support Masters These blackline masters make it easier for students to record their work on selected activities in the Student Edition.

Alternative Lesson Openers An engaging alternative for starting each lesson is provided from among these four types: *Application, Activity, Graphing Calculator,* or *Visual Approach.* (Color transparencies also available)

Graphing Calculator Activities with Keystrokes Keystrokes for four models of calculators are provided for each Technology Activity in the Student Edition, along with alternative Graphing Calculator Activities to begin selected lessons.

Practice A, B, and C These exercises offer additional practice for the material in each lesson, including application problems. There are three levels of practice for each lesson: A (basic), B (average), and C (advanced).

Contents

Reteaching with Practice These two pages provide additional instruction, worked-out examples, and practice exercises covering the key concepts and vocabulary in each lesson.

Quick Catch-Up for Absent Students This handy form makes it easy for teachers to let students who have been absent know what to do for homework and which activities or examples were covered in class.

Cooperative Learning Activities These enrichment activities apply the math taught in the lesson in an interesting way that lends itself to group work.

Interdisciplinary Applications/Real-Life Applications Students apply the mathematics covered in each lesson to solve an interesting interdisciplinary or real-life problem.

Math and History Applications This worksheet expands upon the Math and History feature in the Student Edition.

Challenge: Skills and Applications Teachers can use these exercises to enrich or extend each lesson.

Quizzes The quizzes can be used to assess student progress on two or three lessons.

Chapter Review Games and Activities This worksheet offers fun practice at the end of the chapter and provides an alternative way to review the chapter content in preparation for the Chapter Test.

Chapter Tests A, B, and C These are tests that cover the most important skills taught in the chapter. There are three levels of test: A (basic), B (average), and C (advanced).

SAT/ACT Chapter Test This test also covers the most important skills taught in the chapter, but questions are in multiple-choice and quantitative-comparison format. (See *Alternative Assessment* for multi-step problems.)

Alternative Assessment with Rubrics and Math Journal A journal exercise has students write about the mathematics in the chapter. A multi-step problem has students apply a variety of skills from the chapter and explain their reasoning. Solutions and a 4-point rubric are included.

Project with Rubric The project allows students to delve more deeply into a problem that applies the mathematics of the chapter. Teacher's notes and a 4-point rubric are included.

Cumulative Review These practice pages help students maintain skills from the current chapter and preceding chapters.

LESSON 6.1

TEACHING TIP Use concrete examples to *review* the properties of exponents—students should have previously seen them in Algebra 1. Then ask your students to write the general rule for each property using variables. This way students practice building up from concrete examples toward abstract ideas.

COMMON ERROR When using the *power of a product property*, some students forget to raise the coefficient to the power. For example, they might evaluate $(4x^2)^3$ as $4x^6$. Remind students that the base is $4x^2$ so they must also cube 4.

LESSON 6.2

COMMON ERROR When setting up synthetic substitution, students might forget to write zeros for any missing term of the polynomial. Ask your students to write down the polynomial in standard form and have them fill in any missing terms with zero coefficients, such as shown in Example 2 on page 330. Then, they can just take the coefficient of each term to perform the synthetic division.

TEACHING TIP Graph some polynomials where a table of values seems to contradict the *end behavior* described in the Summary on page 331. For example, for $f(x) = x^3 - 6x^2 + 3x + 10$ the table of values will show the function decreasing between $x = 2$ and $x = 3$. Therefore, students might extend this graph so that $f(x) \to -\infty$ as $x \to \infty$. Point out that a table of values only shows a "local picture" of the function. To predict the *end behavior*, students should use the criteria outlined on page 331, even if it means adding new *turning points* to their graphs.

LESSON 6.3

COMMON ERROR Many students subtract only the first term of the second polynomial and add the other terms. Remind students that the minus sign affects all terms of the second polynomial.

TEACHING TIP You might want to help your students memorize the *square and cube of a binomial patterns* by showing them the similarities between them. Point out that the powers of *a* are written in increasing order, whereas the powers of *b* are in decreasing order. The coefficients are symmetric

and start and end with a 1. Furthermore, the exponent of the binomial becomes a coefficient when it is expanded. Finally, a sum results in all terms being added whereas a difference yields alternate sums and differences. If your students still have trouble using the *square and cube of a binomial patterns*, remind them that they can always use the distributive property to multiply the polynomials.

LESSON 6.4

TEACHING TIP Ask students to write down and memorize the squares and cubes of the first twelve whole numbers. This will help them to identify patterns when they are factoring.

TEACHING TIP When using *grouping* to factor a polynomial, remark that the expression inside the parentheses obtained after factoring the first and the last pairs of terms must be the same. If these expressions are not the same, either a mistake was made, a different monomial must be factored from one of those pairs of terms, or the expression cannot be factored using grouping.

COMMON ERROR Some students will stop factoring too soon. Remind students that they must examine the resulting factors looking for any expressions they might be able to factor again.

LESSON 6.5

COMMON ERROR Some students will be confused by the similarities between *synthetic evaluation* and *synthetic division* and they might not know what sign to use for *k*. Remind students that for *synthetic division* they must write the *opposite of the constant term* of the divisor.

TEACHING TIP Start the lesson by reviewing how to use *long division* of whole numbers. Also review key vocabulary words such as *dividend*, *divisor*, *quotient*, and *remainder*. You might want to write down the relationship between these terms as

$$\frac{\text{Dividend}}{\text{Divisor}} = \text{Quotient} + \frac{\text{Remainder}}{\text{Divisor}}$$

or Dividend = Divisor · Quotient + Remainder to help students write and/or check their answers when they work with polynomials.

LESSON 6.6

COMMON ERROR Ask your students to make a list of the possible rational zeros of a polynomial function, making sure to eliminate "duplicates," such as $\frac{6}{5}$ and $\frac{12}{10}$. Also, remind your students that each time they find a zero of the polynomial function, they must rewrite the subject function as a product of a binomial and another polynomial. This new polynomial is the quotient from the synthetic division, which is the one they must use to continue searching for zeros, *not the original function*.

LESSON 6.7

TEACHING TIP Show your students that giving the zeros of a polynomial function is not enough information to determine a unique answer. Students should understand why they are asked to find a function of *least degree with leading coefficient of 1*. You might want to challenge your students to find other polynomial functions with the same zeros.

TEACHING TIP You might want to create separate forms of assessment with and without graphing calculators and give *both* forms to your students, to make sure they are able to solve problems either way. Otherwise they might always use their graphing calculator to *approximate* the zeros of a polynomial function, without showing understanding of any of the theorems covered in this chapter.

LESSON 6.8

TEACHING TIP Show your students how to identify *repeated solutions* from the graph of a polynomial function. They should be able to distinguish a "double" solution—which also happens to be a *turning point*—from a "single" one by examining the shape of the graph when it crosses the *x*-axis. You might even want to show them how to identify "triple" solutions.

COMMON ERROR Some students erroneously believe that a polynomial function of degree n has *exactly* $n - 1$ turning points. Remind students that this is only true if the function also has n real zeros. Find the turning points for a function such as $f(x) = x^4 - 3x^3 + 2x^2 + 2x - 4$, which has degree 4 but only 2 turning points.

LESSON 6.9

TEACHING TIP Students now know how to model data using either *linear*, *quadratic*, *cubic*, or even *quartic regression*. Have students model a given set of data using each of these forms of regression. Then, students can use the models to predict a value and they can compare how different the values are that they get with each model. You can also teach your students how to use the correlation coefficient, r, and the coefficient of determination, r^2 or R^2, to decide which model best fits the data.

Outside Resources

BOOKS/PERIODICALS

Metz, James R., and Joseph T. Zilliox. "Reciprocal Mappings: The Neglected Transformations." *Mathematics Teacher* (April 1997), pp. 322–327.

Bosch, William, Jennifer Sizoo, Anita Curtis, Shannon Klein, Cheryl Micale, and E-Sin Lin. "Fishy Formulas." *Mathematics Teacher* (November, 1997), pp. 666–671.

SOFTWARE

Harvey, Wayne, and Judah L. Schwartz. *The Function Supposer: Explorations in Algebra*. Newton, MA: Educational Development Center, 1992.

VIDEOS

Apostel, Tom. *Polynomials*. Reston, VA: NCTM.

Parent Guide for Student Success

For use with Chapter 6

Chapter Overview One way that you can help your student succeed in Chapter 6 is by discussing the lesson goals in the chart below. When a lesson is completed, ask your student to interpret the lesson goals for you and to explain how the mathematics of the lesson relates to one of the key applications listed in the chart.

Lesson Title	Lesson Goals	Key Applications
6.1: Using Properties of Exponents	Evaluate and simplify expressions involving powers. Use exponents and scientific notation to solve problems.	• Astronomy • Economics • Ornithology
6.2: Evaluating and Graphing Polynomial Functions	Evaluate and graph a polynomial function.	• Photography • Nursing • Tennis
6.3: Adding, Subtracting, and Multiplying Polynomials	Add, subtract, and multiply polynomials and use polynomial operations in real-life problems.	• Farming • Publishing • Bicycling
6.4: Factoring and Solving Polynomial Equations	Factor polynomial expressions and use factoring to solve polynomial equations.	• Archaeology • City Park • Sculpture
6.5: The Remainder and Factor Theorems	Divide polynomials and use polynomial division in real-life problems.	• Accounting • Fuel Consumption • Movies
6.6: Finding Rational Zeros	Find the rational zeros of a polynomial function. Use polynomial equations to solve real-life problems.	• Crafts • Health Product Sales • Molten Glass
6.7: Using the Fundamental Theorem of Algebra	Determine the number of zeros of a polynomial function and approximate the real zeros.	• Physiology • Education Donations • Television
6.8: Analyzing Graphs of Polynomial Functions	Analyze the graph of a polynomial function to answer questions about real-life situations.	• Manufacturing • Swimming • Quonset Huts
6.9: Modeling with Polynomials	Use finite differences to find the degree of a polynomial that will fit a set of data and find polynomial models for real-life data.	• Boating • Girl Scouts • Space Exploration

Study Strategy

Making a Flow Chart is the study strategy featured in Chapter 6 (see page 322). To help your student understand this strategy, you may wish to have him or her make a flow chart of some household task such as doing laundry. Encourage your student to create and share with you the flow chart described on page 322, since this will be helpful in reviewing the chapter.

Algebra 2
Chapter 6 Resource Book

Parent Guide for Student Success

For use with Chapter 6

Key Ideas Your student can demonstrate understanding of key concepts by working through the following exercises with you.

Lesson	Exercise
6.1	An astronomical unit is 92,960,000 miles, the mean distance of the Earth from the sun. Neptune's farthest distance from the sun is 2,818,000,000 miles. About how many astronomical units is this distance?
6.2	Describe the end behavior of the graph of the polynomial function $f(x) = -x^5 + 3x + 2$ by completing the statements $f(x) \to$? as $x \to -\infty$ and $f(x) \to$? as $x \to \infty$.
6.3	Find the volume V of a box with width w, length $w + 7$, and height $w + 3$ as a polynomial in standard form.
6.4	Solve $x^3 + 5x^2 - 9x - 45 = 0$.
6.5	Factor $f(x) = 3x^3 + 5x^2 - 42x + 40$, given that $f(-5) = 0$.
6.6	You are designing a box to hold 192 cubic centimeters of fruit juice. The length of the box will be 2 centimeters less than the height, and the width of the box will be 4 centimeters less than the height. What should the dimensions of your fruit juice box be?
6.7	Write a polynomial function of least degree with real coefficients, zeros at -4, 3, and $2i$, and a leading coefficient of 1.
6.8	Use a graph to identify the x-intercepts, local maximums, and local minimums of the graph of $f(x) = x^4 - 10x^2 + 9$.
6.9	Write a cubic function whose graph passes through the points $(-1, 0)$, $(0, 0)$, $(2, 0)$, and $(1, -10)$.

Home Involvement Activity

You Will Need: a ruler, a rectangular piece of cardboard

Directions: If you cut a square flap in each corner of the piece of cardboard and fold up the sides, you have a box with an open top. Measure each side of the piece of cardboard. Let x equal the side length of the corner flap. Write an expression for the volume of the box as a function of x. Find the volume of the box for three different values of x. If you have a graphing calculator, graph your volume function to find the value of x that maximizes the volume of the box.

Answers

6.1: about 30 AU **6.2:** $f(x) \to \infty$ as $x \to -\infty$ and $f(x) \to -\infty$ as $x \to \infty$.
6.3: $V = w^3 + 10w^2 + 21w$ **6.4:** 3, -3, -5 **6.5:** $(x + 5)(3x - 4)(x - 2)$ **6.6:** 6 cm long by 4 cm wide by 8 cm high **6.7:** $x^4 + x^3 - 8x^2 + 4x - 48$ **6.8:** x-intercepts are $-1, 1, -3, 3$; local maximum at $(0, 9)$; local minimums at $(2.24, -16)$ and $(-2.24, -16)$ **6.9:** $f(n) = 5n^3 - 5n^2 - 10n$

NAME _____ DATE _____

Prerequisite Skills Review

For use before Chapter 6

EXAMPLE 1 *Simplifying by Combining Like Terms*

Simplify the expression.

$4(2x - 6) - 9x$

SOLUTION

$$4(2x - 6) - 9x = 8x - 24 - 9x \qquad \text{Distributive property}$$
$$= (8x - 9x) - 24 \qquad \text{Group like terms.}$$
$$= -x - 24 \qquad \text{Combine like terms.}$$

Exercises for Example 1

Simplify the expression.

1. $-3(3x + 1) + 5x$

2. $-x^2 + 3x^3 + 4x^2 - 6x^3$

3. $-3x^4 + 6x^2 - 8x^2 + 2x^3$

4. $13x^2 - 6(3x^2 + 2)$

EXAMPLE 2 *Graphing a Quadratic Function*

Graph the quadratic function.

$y = (x - 1)(x - 5)$

SOLUTION

The quadratic function is in intercept form $y = a(x - p)(x - q)$ where $a = 1, p = 1$, and $q = 5$. The x-intercepts occur at $(1, 0)$ and $(5, 0)$. The axis of symmetry lies halfway between these points, at $x = 3$. So, the x-coordinate of the vertex is $x = 3$ and the y-coordinate of the vertex is:

$y = (3 - 1)(3 - 5) = -4.$

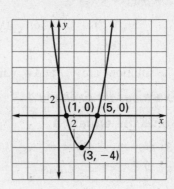

Exercises for Example 2

Graph the quadratic function.

5. $y = (x + 3)(x - 5)$

6. $y = -(x + 3)^2$

7. $y = -2(x - 3)(x + 1)$

8. $y = (x - 1)^2$

NAME _____ DATE _____

Prerequisite Skills Review

For use before Chapter 6

EXAMPLE 3 *Writing Quadratic Functions in Standard Form*

Write the quadratic function in standard form.

$$y = (x + 7)^2 - 8$$

SOLUTION

$y = (x + 7)^2 - 8$	Write original function.
$\quad = (x + 7)(x + 7) - 8$	Rewrite $(x + 7)^2$.
$\quad = (x^2 + 7x + 7x + 49) - 8$	Multiply using FOIL.
$\quad = x^2 + 14x + 41$	Combine like terms.

Exercises for Example 3

Write the quadratic function in standard form.

9. $y = (x - 4)^2 - 3$ **10.** $y = -(x - 9)(x - 6)$

11. $y = -(x - 5)^2 + 1$ **12.** $y = 3(x + 1)(x - 2)$

EXAMPLE 4 *Solving Quadratic Equations*

Solve the equation.

$$x^2 + 3x - 28 = 0$$

SOLUTION

$x^2 + 3x - 28 = 0$	Write original equation.
$(x + 7)(x - 4) = 0$	Factor.
$(x + 7) = 0 \quad$ or $\quad (x - 4) = 0$	Zero product property
$x = -7 \quad$ or $\quad x = 4$	Solve for x.

The solutions are -7 and 4. Check the solutions in the original equation.

Exercises for Example 4

Solve the equation.

13. $x^2 - 5x - 24 = 0$ **14.** $2x^2 - 13x - 7 = 0$

15. $x^2 - 15x + 44 = 0$ **16.** $x^2 + 8x - 20 = 0$

Strategies for Reading Mathematics

For use with Chapter 6

Strategy: Reading Standard Form

Each type of relation has a standard form. Often the standard form is the one most useful for identifying the type of relation and for graphing it, finding its roots, and so on. A polynomial function is equal to a sum of terms of the type $a_k x^k$, where k is a whole number. A polynomial function is in standard form if its terms are written in decreasing order of exponents from left to right. When in standard form, the exponent of the first term is the degree of the polynomial function and its coefficient is the leading coefficient.

EXAMPLE

Tell whether the function is a polynomial function. If it is, write it in standard form and identify the degree and leading coefficient.

a. $f(x) = -5x^2 - 7x + x^5 + x^{-1}$ **b.** $f(x) = 3x^3 + x^4 + 20$

SOLUTION

a. This function is not a polynomial since the exponent of one term is -1, which is not a whole number.

b. This function is a polynomial function. In standard form, it is written as $f(x) = x^4 + 3x^3 + 20$. It is a fourth degree polynomial, and the leading coefficient is 1.

> **STUDY TIP**
> **Reading Standard Form**
> Learn the standard form of each type of function or relation as it is introduced. To see if an equation is a particular type of relation, it may help to try to put it in the appropriate standard form.

Questions

1. Tell why each function is not a polynomial function in standard form.

    a. $f(x) = 3x^4 + 2x^2 - 6x + \dfrac{1}{x^2}$ **b.** $f(x) = 4x^2 - 6x^2 + 3x - 7$

    c. $f(x) = 18x + 7x^3 - x^5$ **d.** $f(x) = 10 + x^5 - 2x^3 + 5x$

2. Give the standard form for each polynomial function, and give its degree and leading coefficient.

    a. $f(x) = 12 - 3x + x^2 + x^3$ **b.** $f(x) = -7x^2 - 3x + 6x^3 + 6 - x^7$

3. To apply the quadratic formula, a quadratic equation must be put in standard form, $ax^2 + bx + c = 0$. Put each quadratic equation in standard form, and identify the values a, b, and c.

    a. $3x^2 = 2x + 7$ **b.** $4x - x^2 = x^2 + 6$ **c.** $x(x + 5) = 14$

Strategies for Reading Mathematics
For use with Chapter 6

Visual Glossary

The Study Guide on page 322 lists the key vocabulary for Chapter 6 as well
as review vocabulary from previous chapters. Use the page references on
page 322 or the Glossary in the textbook to review key terms from prior
chapters. Use the visual glossary below to help you understand some of the
key vocabulary in Chapter 6. You may want to copy these diagrams into your
notebook and refer to them as you complete the chapter.

GLOSSARY

Polynomial function (p. 329)
A function of the form
$f(x) = a_n x^n + a_{n-1} x^{n-1} + \cdots + a_1 x + a_0$ where $a_n \neq 0$
and the exponents are all
whole numbers.

End behavior (p. 331) The
behavior of the graph of a
function as x approaches
positive infinity or negative
infinity.

Local maximum (p. 374)
The y-coordinate of the
turning point of a graph of a
function if the point is higher
than all nearby points.

Local minimum (p. 374)
The y-coordinate of the
turning point of a graph of a
function if the point is lower
than all nearby points.

Synthetic division (p. 353)
A method used to divide a
polynomial by an expression
of the form $x - k$.

Graphs of Polynomial Functions

The domain of a polynomial function is all real numbers, and its graph
is a smooth curve. The degree and coefficients of the function
determine such things as its end behavior and general shape.

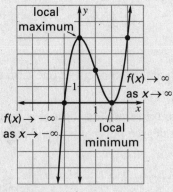

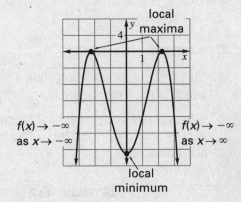

$$f(x) = (x + 1)(x - 2)^2$$
$$= x^3 - 3x^2 + 4$$

Synthetic Division

You can use synthetic division to find the value of a polynomial
function at a specific x-value, or to find roots of the factored form
of a polynomial.

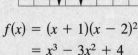

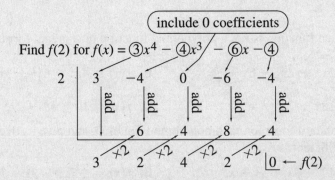

Since $f(2) = 0$, 2 is a root of $f(x)$.

Algebra 2
Chapter 6 Resource Book

TEACHER'S NAME _____ CLASS _____ ROOM _____ DATE _____

Lesson Plan

1-day lesson (See *Pacing the Chapter*, TE pages 320C–320D)   For use with pages 323–328

 GOALS
1. **Use properties of exponents to evaluate and simplify expressions involving powers.**
2. **Use exponents and scientific notation to solve real-life problems.**

State/Local Objectives _____

_____

✓ Check the items you wish to use for this lesson.

STARTING OPTIONS
____ Prerequisite Skills Review: CRB pages 5–6
____ Strategies for Reading Mathematics: CRB pages 7–8
____ Warm-Up or Daily Homework Quiz: TE pages 323 and 311, CRB page 11, or Transparencies

TEACHING OPTIONS
____ Motivating the Lesson: TE page 324
____ Lesson Opener (Application): CRB page 12 or Transparencies
____ Examples 1–4: SE pages 324–325
____ Extra Examples: TE pages 324–325 or Transparencies; Internet
____ Closure Question: TE page 325
____ Guided Practice Exercises: SE page 326

APPLY/HOMEWORK
Homework Assignment
____ Basic   16–46 even, 49–53 odd, 57, 61–81 odd
____ Average   16–46 even, 48–53, 57, 61–81 odd
____ Advanced   16–46 even, 48–59, 61–81 odd

Reteaching the Lesson
____ Practice Masters: CRB pages 13–15 (Level A, Level B, Level C)
____ Reteaching with Practice: CRB pages 16–17 or Practice Workbook with Examples
____ Personal Student Tutor

Extending the Lesson
____ Applications (Real-Life): CRB page 19
____ Challenge: SE page 328; CRB page 20 or Internet

ASSESSMENT OPTIONS
____ Checkpoint Exercises: TE pages 324–325 or Transparencies
____ Daily Homework Quiz (6.1): TE page 328, CRB page 23, or Transparencies
____ Standardized Test Practice: SE page 328; TE page 328; STP Workbook; Transparencies

Notes _____

_____

_____

TEACHER'S NAME _____ CLASS _____ ROOM _____ DATE _____

Lesson Plan for Block Scheduling

Half-day lesson (See *Pacing the Chapter,* TE pages 320C–320D)   For use with pages 323–328

GOALS   1. **Use properties of exponents to evaluate and simplify expressions involving powers.**
2. **Use exponents and scientific notation to solve real-life problems.**

State/Local Objectives _____

_____

_____

✓ **Check the items you wish to use for this lesson.**

STARTING OPTIONS

____ Prerequisite Skills Review: CRB pages 5–6
____ Strategies for Reading Mathematics: CRB pages 7–8
____ Warm-Up or Daily Homework Quiz: TE pages 323 and 311,
          CRB page 11, or Transparencies

TEACHING OPTIONS

____ Motivating the Lesson: TE page 324
____ Lesson Opener (Application): CRB page 12 or Transparencies
____ Examples 1–4: SE pages 324–325
____ Extra Examples: TE pages 324–325 or Transparencies; Internet
____ Closure Question: TE page 325
____ Guided Practice Exercises: SE page 326

APPLY/HOMEWORK

Homework Assignment (See also the assignment for Lesson 6.2.)

____ Block Schedule: 16–46 even, 48–53, 57, 61–81 odd

Reteaching the Lesson

____ Practice Masters: CRB pages 13–15 (Level A, Level B, Level C)
____ Reteaching with Practice: CRB pages 16–17 or Practice Workbook with Examples
____ Personal Student Tutor

Extending the Lesson

____ Applications (Real Life): CRB page 19
____ Challenge: SE page 328; CRB page 20 or Internet

ASSESSMENT OPTIONS

____ Checkpoint Exercises: TE pages 324–325 or Transparencies
____ Daily Homework Quiz (6.1): TE page 328, CRB page 23, or Transparencies
____ Standardized Test Practice: SE page 328; TE page 328; STP Workbook; Transparencies

CHAPTER PACING GUIDE	
Day	Lesson
1	**6.1 (all)**; 6.2(all)
2	6.3 (all)
3	6.4 (all)
4	6.5 (all); 6.6(all)
5	6.7 (all); 6.8 (all)
6	6.9(all); Review Ch. 6
7	Assess Ch. 6; 7.1 (all)

Notes _____

_____

_____

WARM-UP EXERCISES

For use before Lesson 6.1, pages 323–328

Evaluate each expression.

1. 2^4

2. $(-2)^4$

3. -2^4

4. 2^3

5. $\left(\frac{1}{2}\right)^5$

DAILY HOMEWORK QUIZ

For use after Lesson 5.8, pages 306–312

1. Write a quadratic function in vertex form for the parabola shown.

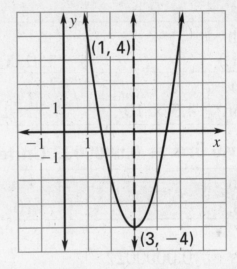

2. Write a quadratic function in intercept form for the parabola shown.

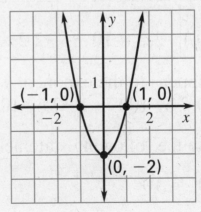

3. Write a quadratic function in standard form for the parabola whose graph passes through $(2, -2)$, $(3, 4)$, and $(0, -2)$.

NAME _____ DATE _____

Application Lesson Opener

For use with pages 323–328

1. Use the exchange rate above each table to find the values in the second row.

2. Complete the third row of each table by rewriting the values from the second row in scientific notation. A number is expressed in *scientific notation* if it is in the form $c \times 10^n$, where $1 \le c < 10$ and n is an integer.

1 United States dollar = 450,000 Turkish lira (TL)

Dollars	0.1	1	10	100
Turkish lira (TL)		450,000		
TL (sci. notation)	$4.5 \times 10^{\square}$	4.5×10^5	$4.5 \times 10^{\square}$	$4.5 \times 10^{\square}$

1 Turkish lira = 0.0000022 Unites States dollar

Turkish lira (TL)	1	10	100	1000
Dollars	0.0000022			
Dollars (sci. not.)	2.2×10^{-6}	$2.2 \times 10^{\square}$	$2.2 \times 10^{\square}$	$2.2 \times 10^{\square}$

3. What happens to the exponent when you multiply a power of ten by $10 = 10^1$? by $100 = 10^2$? by $1000 = 10^3$? by $0.1 = 10^{-1}$?

4. Complete the following statement for integers n and k:

$$10^n \cdot 10^k = 10^{\square}$$

5. Extend the patterns in the tables to find the values.

   a. 1¢ = _____ TL   **b**. _____ TL = $2.20

NAME _____ DATE _____

Practice A

For use with pages 323–328

Use the properties of exponents to evaluate the expression.

1. $3^4 \cdot 3^5$

2. $2^6 \cdot 2^2$

3. $4^3 \cdot 4^2$

4. $4^{-1} \cdot 4^4$

5. $2^{-5} \cdot 2^3$

6. $5^{-7} \cdot 5^8$

7. $6^{-2} \cdot 6^{-1}$

8. $3^{-2} \cdot 3^{-3}$

9. $2^{-4} \cdot 2^{-3}$

10. $(-3)^2(-3)^3$

11. $(-5)^{-6}(-5)^8$

12. $(-2)^{-3}(-2)^{-2}$

13. $\dfrac{5^4}{5^2}$

14. $\dfrac{7^6}{7^9}$

15. $\dfrac{3^5}{3^5}$

16. $\dfrac{(-2)^8}{(-2)^3}$

17. $\dfrac{(-3)^3}{(-3)^4}$

18. $(5^2)^3$

19. $(2^4)^2$

20. $(3^2)^3$

21. $\left(\dfrac{1}{3}\right)^4$

22. $\left(\dfrac{3}{2}\right)^3$

23. $\left(-\dfrac{2}{5}\right)^2$

24. $\left(-\dfrac{1}{4}\right)^3$

25. 13^0

26. $\left(\dfrac{4}{5}\right)^0$

27. 3^{-2}

28. 4^{-3}

29. $\left(\dfrac{3}{4}\right)^{-2}$

30. $\left(\dfrac{2}{3}\right)^{-4}$

Simplify the expression.

31. $x^3 \cdot x^5$

32. $x^4 \cdot x^8$

33. $(x^4)^6$

34. $(3x)^3$

35. $\left(\dfrac{x}{2}\right)^4$

36. $\dfrac{x^7}{x^2}$

37. $\dfrac{x^3}{x^9}$

38. $\left(\dfrac{3}{x}\right)^2$

39. $\left(\dfrac{x}{4}\right)^{-2}$

Surface Area   **In Exercises 40–42, use the formula $S = 4\pi r^2$ to find the surface area of each planet.**

40. The radius of Jupiter is approximately 44,366 miles. Find the surface area of Jupiter.

41. The radius of Earth is approximately 3863 miles. Find the surface area of Earth.

42. The radius of Mars is approximately 2110 miles. Find the surface area of Mars.

NAME _____ DATE _____

Practice B

For use with pages 323–328

Use the properties of exponents to evaluate the expression.

1. $(3^4)(3^{-2})$

2. $(5^2)^3$

3. $\left(\dfrac{2}{3}\right)^3$

4. $\dfrac{8^4}{8^6}$

5. $(7^6)(7^{-6})$

6. $\dfrac{4 \cdot 4^3}{4^6}$

7. $\dfrac{(3^2)^5}{3^8}$

8. $\left(\dfrac{1}{2}\right)^{-4}$

9. $\dfrac{5^6}{(5^3)^2}$

Simplify the expression.

10. $x^3 \cdot x^2$

11. $\dfrac{2y^3}{y^5}$

12. $(3x)^2$

13. $\left(\dfrac{y}{2}\right)^3$

14. $(4x^3)^4$

15. $x^0 y^{-2}$

16. $\dfrac{5x^2 y}{2x^{-1}y^3}$

17. $\dfrac{-3xy}{9x^3 y^{-4}}$

18. $\dfrac{(3x)^2}{6x^5}$

19. *Geometry*   Find an expression for the area of the triangle.

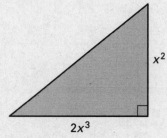

x^2

$2x^3$

20. *Geometry*   Find an expression for the area of the circle.

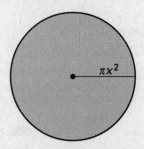

πx^2

21. *Population per Square Mile*   In 1996, the population of the United States was approximately 265,280,000 people. The area of the United States is approximately 3,780,000 square miles. Use scientific notation to find the population per square mile in the United States.

22. *Speed of Mercury*   Mercury travels approximately 226,000,000 miles around the sun. It takes Mercury approximately 2100 hours to revolve around the sun. Use scientific notation to find the speed of Mercury as it revolves around the sun.

23. *Computers per 1000 People*   The population of the United States is approximately 265,280 thousand people. It is estimated that by the year 2000, there will be 154,000,000 computers in the United States. How many computers will there be per 1000 people?

NAME _____ DATE _____

Practice C
For use with pages 323–328

Use the properties of exponents to evaluate the expression.

1. $\dfrac{2^{-4} \cdot 2^5}{2^{-3}}$

2. $\dfrac{(3^2)^3}{3^{-1}}$

3. $\dfrac{(-2)^4(-2)^{-1}}{(-2)^3}$

4. $\left(\dfrac{2}{3}\right)^{-3}$

5. $\left[\left(\dfrac{3}{8}\right)^2\right]^2$

6. $\dfrac{\left(\dfrac{1}{3}\right)^{-3}}{\left(\dfrac{1}{3}\right)^4\left(\dfrac{1}{3}\right)^{-2}}$

Simplify the expression.

7. $\dfrac{xy}{4} \cdot \dfrac{2x^2}{y^3}$

8. $\dfrac{-2x^2}{3xy^3} \cdot \dfrac{2x^{-1}}{y^{-1}}$

9. $\dfrac{x^{-4}}{y^{-2}} \cdot \dfrac{y^{-2}}{x^{-4}}$

10. $\dfrac{5x^2y}{8} \cdot \dfrac{-2x^{-1}y}{x^3y}$

11. $\dfrac{(2x^2)^3}{5} \cdot \dfrac{15x^{-2}}{2x^3}$

12. $\left(\dfrac{x^3y^2}{2x^4}\right)^{-2}$

13. $\left(\dfrac{(2x)^3}{2x^3}\right) \cdot 4x^5$

14. $\left(\dfrac{x^2y^5z}{2x^3}\right)^2$

15. $\left[(x^4)^{-6}\right]^2$

Use the properties of exponents to simplify the left side of the equation. Then solve the equation as demonstrated below.

$$4^{x-1} = 4^2 \implies x - 1 = 2 \implies x = 3$$

16. $2^x2^3 = 2^5$

17. $\dfrac{3^x}{3^2} = 3^4$

18. $(5^x)^3 = 5^{12}$

19. $\dfrac{4^3}{4^x} = 4^0$

20. $\dfrac{2^{-x}y^2}{y^2} = 2^5$

21. $(-2x)^0(3^2)(3^x) = 3^{-1}$

Class Project **In Exercises 22–25, use the following information.**
Your class project is to design a piece of playground equipment for an
elementary school. You design a romper room that will contain small plastic
balls for the children to roll around in. The room will be 10 feet by 10 feet. The
plastic balls will cover the entire floor to a depth of 2 feet. A toy distributor can
ship you 190 balls (each with a radius of $1\frac{3}{4}$ inches) in a cubic box, 20 inches on
a side.

22. Find an expression for the volume (in cubic inches) of one ball.

23. Find an expression that represents the ratio of the volume of 190
balls to the volume of the cubic box.

24. What percent of the volume of the cubic box is filled with plastic
balls?

25. Find the volume of the region in the romper room that will contain
plastic balls. Give your result in cubic inches.

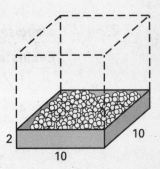

NAME _____ DATE _____

Reteaching with Practice

For use with pages 323–328

GOAL How to use properties of exponents to evaluate and simplify expressions involving powers and to use exponents and scientific notation to solve real-life problems

---

VOCABULARY

The following are properties of exponents.

Let a and b be real numbers and let m and n be integers.

Product of Powers Property	$a^m \cdot a^n = a^{m+n}$
Power of a Power Property	$(a^m)^n = a^{mn}$
Power of a Product Property	$(ab)^m = a^m b^m$
Negative Exponent Property	$a^{-m} = \dfrac{1}{a^m}, a \neq 0$
Zero Exponent Property	$a^0 = 1, a \neq 0$
Quotient of Powers Property	$\dfrac{a^m}{a^n} = a^{m-n}, a \neq 0$
Power of a Quotient Property	$\left(\dfrac{a}{b}\right)^m = \dfrac{a^m}{b^m}, b \neq 0$

A number is expressed in **scientific notation** if it is in the form $c \times 10^n$ where $1 \leq c < 10$ and n is an integer.

EXAMPLE 1 *Evaluating Numerical Expressions*

---

a. $\dfrac{3^6}{3^2} = 3^{6-2}$ Quotient of powers property

   $= 3^4$ Simplify exponent

   $= 81$ Evaluate power.

b. $(-6)(-6)^{-1} = (-6)^{1+(-1)}$ Product of powers property

   $= (-6)^0$ Simplify exponent.

   $= 1$ Zero exponent property

Exercises for Example 1

Evaluate the expression.

1. $(2^3)^2$

2. $8^3 \cdot 8$

3. $(3^2)^4$

4. $\left(\dfrac{2}{3}\right)^3$

5. $\left(\dfrac{1}{5}\right)^{-2}$

6. $\dfrac{3^3}{3^2}$

NAME _____ DATE _____

Reteaching with Practice

For use with pages 323–328

EXAMPLE 2 *Simplifying Algebraic Expressions*

Remember that simplified algebraic expressions contain only positive exponents.

a. $\dfrac{x^7 y^4}{x^{-1} y^{-2}} = x^{7-(-1)} y^{4-(-2)}$ Quotient of powers property

$\qquad\qquad = x^8 y^6$ Simplify exponents.

b. $(3x^{-2} y^4)^2 = 3^2 (x^{-2})^2 (y^4)^2$ Power of a product property

$\qquad\qquad = 9x^{-4} y^8$ Power of a power property

$\qquad\qquad = \dfrac{9y^8}{x^4}$ Negative exponent property

c. $(-2x^5 y^3)^{-2} = (-2)^{-2} (x^5)^{-2} (y^3)^{-2}$ Power of a product property

$\qquad\qquad = (-2)^{-2} x^{-10} y^{-6}$ Power of a power property

$\qquad\qquad = \dfrac{1}{4x^{10} y^6}$ Negative exponent property

Exercises for Example 2

Simplify the expression.

7. $(2x^2)^5$ **8.** $\dfrac{y^{-2}}{y^3}$ **9.** $3^3 \cdot 3^{-5}$ **10.** $(x^2 y^4)^{-3}$ **11.** $\dfrac{x^4 y^{10}}{xy^3}$

EXAMPLE 3 *Writing Numbers in Scientific Notation*

Alaska is the largest state in the United States, with an area of 1,530,693 square kilometers. Express Alaska's area in scientific notation.

SOLUTION

Recall that a number expressed in scientific notation has the form $c \times 10^n$.

1.530693×10^6 Place the decimal point between the 1 and 5 since $1 \le c < 10$.

The power of 10 is 6 since the decimal point was shifted 6 places to the left.

The area of Alaska is about 1.5 million square kilometers.

Exercises for Example 3

12. On average, the planet Pluto's distance from the sun is about 3,670,000,000 miles. Express the distance in scientific notation.

13. The Pacific Ocean is the world's largest ocean, covering 70,000,000 square miles of the Earth's surface. Express this area in scientific notation.

NAME _____ DATE _____

Quick Catch-Up for Absent Students

For use with pages 323–328

The items checked below were covered in class on (date missed) _____

Lesson 6.1: Using Properties of Exponents

____ **Goal 1:** Use properties of exponents to evaluate and simplify expressions involving powers.
     (pp. 323, 324)

Material Covered:

    ____ Activity: Products and Quotients of Powers

    ____ Student Help: Study Tip

    ____ Example 1: Evaluating Numerical Expressions

    ____ Example 2: Simplifying Algebraic Expressions

____ **Goal 2:** Use exponents and scientific notation to solve real-life problems. (p. 325)

Material Covered:

    ____ Example 3: Comparing Real-Life Volumes

    ____ Student Help: Skills Review

    ____ Example 4: Using Scientific Notation in Real-Life

Vocabulary:

    scientific notation, p. 325

____ Other (specify) _____

_____

Homework and Additional Learning Support

    ____ Textbook (specify) _pp. 326–328_____

    _____

    ____ Internet: Extra Examples at www.mcdougallittell.com

    ____ *Reteaching with Practice* worksheet (specify exercises)_____

    ____ *Personal Student Tutor* for Lesson 6.1

NAME _____ DATE _____

Real-Life Application:
When Will I Ever Use This?

For use with pages 323–328

Planetary Motion

Johannes Kepler (1571–1630) discovered that the orbits of planets in our solar system follow paths around the sun that are ellipses. Kepler discovered several laws about planetary motion, one of which is described below.

In Exercises 1–6, use the following information.

In 1619, Kepler discovered that the period, P (in years, where one year is 365.25 days), of each planet in our solar system is related to the planet's mean distance, a (in astronomical units), from the sun by the equation $\frac{P^2}{a^3} = k$. This relationship is often referred to as Kepler's third law of planetary motion.

1. One astronomical unit is approximately (a) 149,600,000 kilometers or (b) 93,000,000 miles. Write both numbers by using scientific notation.

2. Test Kepler's equation for the nine planets in our solar system, using the table at the right. (Astronomical units relate the other planets' periods and mean distances to Earth's period and mean distance.) Is k approximately the same value for each planet?

Planet	P	a
Mercury	0.241	0.387
Venus	0.615	0.723
Earth	1.000	1.000
Mars	1.881	1.523
Jupiter	11.861	5.203
Saturn	29.457	9.541
Uranus	84.008	19.190
Neptune	164.784	30.086
Pluto	248.350	39.507

3. When looking through your telescope you discover a planet whose mean distance from the sun is 48.125 astronomical units. Use the results in Exercise 1 to find the period of this planet in days. (The period of Earth is 365.25 days.)

4. Find the ratio of Pluto's mean distance (from the sun) to Mercury's mean distance.

5. You are drawing a diagram of our solar system in which Mercury's mean distance from the sun is represented by 1 inch. How many inches will be needed to represent Pluto's mean distance from the sun?

6. Redo the table given in Exercise 1 using scientific notation.

NAME _____ DATE _____

Challenge: Skills and Applications

For use with pages 323–328

1. Show that for any positive integer m, $2^m + 2^m = 2^{m+1}$.

2. Show that for any positive integer n, $2^n + 2^{n+1} = 3 \cdot 2^n$.

3. Pierre Fermat (1601–1665) conjectured that every number F_n of the form $2^{2^n} + 1$, where n is a nonnegative integer, is prime. These numbers are called *Fermat numbers* in his honor.

   a. Find the values of F_n for $n = 0, 1, 2, 3, 4$.

   b. Use a calculator to find the value of F_5. In about 1750, another mathematician, Leonhard Euler, showed that this number is the product of 641 and another integer (both of which are prime). Find this other integer.

   c. Even though Fermat's conjecture turned out to be wrong, it was later shown that a regular polygon with n sides could be constructed perfectly with a compass and a straightedge if and only if $n = 2^r \cdot F$, where r is a nonnegative integer (possibly 0) and F is a Fermat number that *is* prime. Tell whether regular polygons with the following numbers of sides can be constructed with a compass and a straightedge: 3, 7, 9, 12, 15, 20, 34, 48, 257.

4. The following formula gives the monthly cost M of a mortgage (a loan taken out with real estate as collateral) if the principal (the amount borrowed) is p dollars, which is to be totally repaid in n months, and the monthly interest rate (as a decimal) is r:

   $$M = \frac{rp}{1 - (1 + r)^{-n}}$$

   a. Simplify this formula.

   b. Find the monthly cost of a mortgage with a principal of $120,000 and a monthly rate of 0.8%, to be totally repaid in 25 years.

5. **a.** Suppose n is an integer. Evaluate $(-1)^{2n}$ and $(-1)^{2n+1}$.

   b. Express i^{2n} and i^{2n+1} in $a + bi$ form, in terms of n. (*Hint:* Use the Properties of Exponents.)

   c. Repeat part (b) for i^{-2n} and $i^{-(2n+1)}$.

Lesson Plan

1-day lesson (See *Pacing the Chapter*, TE pages 320C–320D) For use with pages 329–337

 GOALS
1. **Evaluate a polynomial function.**
2. **Graph a polynomial function.**

State/Local Objectives _____

_____

✓ Check the items you wish to use for this lesson.

STARTING OPTIONS
_____ Homework Check: TE page 326 Answer Transparencies
_____ Warm-Up or Daily Homework Quiz: TE pages 329 and 328, CRB page 23, or Transparencies

TEACHING OPTIONS
_____ Motivating the Lesson: TE page 330
_____ Lesson Opener (Visual Approach): CRB page 24 or Transparencies
_____ Graphing Calculator Activity with Keystrokes: CRB page 25
_____ Examples 1–5: SE pages 329–332
_____ Extra Examples: TE pages 330–332 or Transparencies; Internet
_____ Technology Activity: SE page 337
_____ Closure Question: TE page 332
_____ Guided Practice Exercises: SE page 333

APPLY/HOMEWORK
Homework Assignment
_____ Basic   16–32 even, 38–44 even, 50–60 even, 65–71 odd, 81, 83, 87, 91–107 odd
_____ Average   16–58 even, 65–73 odd, 80, 82–84, 87, 91–107 odd
_____ Advanced   16–60 even, 65–75 odd, 80–89, 91–107 odd

Reteaching the Lesson
_____ Practice Masters: CRB pages 26–28 (Level A, Level B, Level C)
_____ Reteaching with Practice: CRB pages 29–30 or Practice Workbook with Examples
_____ Personal Student Tutor

Extending the Lesson
_____ Applications (Interdisciplinary): CRB page 32
_____ Challenge: SE page 336; CRB page 33 or Internet

ASSESSMENT OPTIONS
_____ Checkpoint Exercises: TE pages 330–332 or Transparencies
_____ Daily Homework Quiz (6.2): TE page 336, CRB page 36, or Transparencies
_____ Standardized Test Practice: SE page 336; TE page 336; STP Workbook; Transparencies

Notes _____

_____

_____

TEACHER'S NAME _____ CLASS _____ ROOM _____ DATE _____

Lesson Plan for Block Scheduling

Half-day lesson (See *Pacing the Chapter,* TE pages 320C–320D)   **For use with pages 329–337**

GOALS **1. Evaluate a polynomial function.**
2. Graph a polynomial function.

State/Local Objectives _____

_____

_____

CHAPTER PACING GUIDE	
Day	**Lesson**
1	6.1 (all); **6.2(all)**
2	6.3 (all)
3	6.4 (all)
4	6.5 (all); 6.6(all)
5	6.7 (all); 6.8 (all)
6	6.9(all); Review Ch. 6
7	Assess Ch. 6; 7.1 (all)

✓ **Check the items you wish to use for this lesson.**

STARTING OPTIONS

____ Homework Check: TE page 326 Answer Transparencies
____ Warm-Up or Daily Homework Quiz: TE pages 329 and 328,
     CRB page 23, or Transparencies

TEACHING OPTIONS

____ Motivating the Lesson: TE page 330
____ Lesson Opener (Visual Approach): CRB page 24 or Transparencies
____ Graphing Calculator Activity with Keystrokes: CRB page 25
____ Examples 1–5: SE pages 329–332
____ Extra Examples: TE pages 330–332 or Transparencies; Internet
____ Technology Activity: SE page 337
____ Closure Question: TE page 332
____ Guided Practice Exercises: SE page 333

APPLY/HOMEWORK

Homework Assignment (See also the assignment for Lesson 6.1.)

____ Block Schedule: 16–58 even, 65–73 odd, 80, 82–84, 87, 91–107 odd

Reteaching the Lesson

____ Practice Masters: CRB pages 26–28 (Level A, Level B, Level C)
____ Reteaching with Practice: CRB pages 29–30 or Practice Workbook with Examples
____ Personal Student Tutor

Extending the Lesson

____ Applications (Interdisciplinary): CRB page 32
____ Challenge: SE page 336; CRB page 33 or Internet

ASSESSMENT OPTIONS

____ Checkpoint Exercises: TE pages 330–332 or Transparencies
____ Daily Homework Quiz (6.2): TE page 336, CRB page 36, or Transparencies
____ Standardized Test Practice: SE page 336; TE page 336; STP Workbook; Transparencies

Notes _____

_____

_____

Lesson 6.2

WARM-UP EXERCISES

For use before Lesson 6.2, pages 329–337

Identify each function as linear or quadratic.

1. $f(x) = 2x^2 + x - 6$

2. $f(x) = 5x + 3$

Find $f(x)$ when $x = -2$.

3. $f(x) = 2x - 9$

4. $f(x) = x^2 - 5x + 7$

5. $f(x) = 3x^3 + 10$

DAILY HOMEWORK QUIZ

For use after Lesson 6.1, pages 323–328

Evaluate the expression.

1. $(-2)^{-3}(-2)^7$

2. $\dfrac{7}{7^{-2}} \cdot 3^5 \cdot 3^{-5}$

Simplify the expression.

3. $\dfrac{4x^{-2}y^3}{32x^5y^{-6}}$

4. $\dfrac{28x^4y^2}{7xy^5}$

5. Write an expression for the volume of the cone in terms of x.

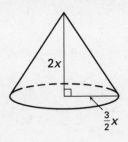

A function $f(x)$ is a *polynomial function* if it can be written as a sum of terms, where each term is a constant times a whole-number power of x. Note that constant terms can occur in polynomial functions because, for example, $7 = 7x^0$.

These are polynomial functions:

$$f(x) = 5x + 17 \qquad\qquad f(x) = 3x^5 - 2x^3 + 11x$$

$$f(x) = x^2 + 4x - 7 \qquad\qquad f(x) = 5x^{16}$$

$$f(x) = \frac{3}{2}x^4 - \frac{2}{3}x^2 \qquad\qquad f(x) = \sqrt{3}\,x^3 + \sqrt{2}\,x^2 + x$$

These are *not* polynomial functions:

$$f(x) = \sqrt{x + 5} \qquad\qquad f(x) = 2x^{-3} + 3x^{-4} + 4x^{-5}$$

$$f(x) = \frac{2x + 3}{5x - 2} \qquad\qquad f(x) = \frac{1}{x^2 - 5x + 3}$$

Decide whether the function is a polynomial function.

1. $f(x) = \sqrt{x - 3}$

2. $f(x) = 5x^4 - 3x^2$

3. $f(x) = 15x^2 + 5\sqrt{3x - 2}$

4. $f(x) = x^3 + 3^x$

5. $f(x) = \dfrac{3x - 5}{2x^3 + 5x - 4}$

6. $f(x) = \dfrac{x^3 - 4x^2 + 3x + 2}{5}$

7. $f(x) = 125x^5 - 32x^4 + 12x^3 - x^2 + 16x + 18$

8. $f(x) = 13x^5 + 12x^4 - \dfrac{2}{3}x^3 + \sqrt{21}\,x^2 + 5x - 2$

9. $f(x) = \sqrt{6x^3 - 12x^3 + 17x + 24}$

Graphing Calculator Activity Keystrokes
For use with page 337

TI-82

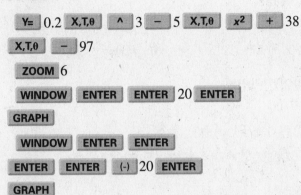

Y= 0.2 X,T,θ ^ 3 − 5 X,T,θ x² + 38
X,T,θ − 97
ZOOM 6
WINDOW ENTER ENTER 20 ENTER
GRAPH
WINDOW ENTER ENTER
ENTER ENTER (-) 20 ENTER
GRAPH

TI-83

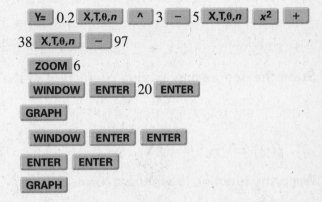

Y= 0.2 X,T,θ,n ^ 3 − 5 X,T,θ,n x² +
38 X,T,θ,n − 97
ZOOM 6
WINDOW ENTER 20 ENTER
GRAPH
WINDOW ENTER ENTER
ENTER ENTER
GRAPH

SHARP EL-9600c

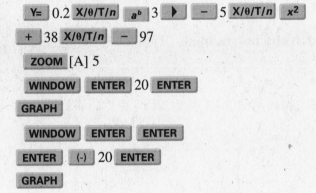

Y= 0.2 X/θ/T/n aᵇ 3 ▶ − 5 X/θ/T/n x²
+ 38 X/θ/T/n − 97
ZOOM [A] 5
WINDOW ENTER 20 ENTER
GRAPH
WINDOW ENTER ENTER
ENTER (-) 20 ENTER
GRAPH

CASIO CFX-9850Gᴀ PLUS

From the main menu, choose GRAPH.

0.2 X,θ,T ^ 3 − 5 X,θ,T x² + 38
X,θ,T − 97 EXE
SHIFT F3 F3 EXIT F6
F3 (-) 10 EXE 20 EXE 1 EXE (-)
10 EXE 10 EXE 1 EXE EXIT F6
F3 (-) 10 EXE 20 EXE 1 EXE (-)
20 EXE 10 EXE 1 EXE EXIT F6

Practice A

For use with pages 329–336

State whether the following function is a polynomial.

1. $f(x) = 3x^2 + 7x - 3$

2. $f(x) = 5 - 3x^4$

3. $f(x) = 2^x - 3x + 1$

4. $f(x) = 9$

5. $f(x) = 2\sqrt{x} + 5x - 8$

6. $f(x) = x\sqrt{3} + x^2 - \pi$

State the degree and leading coefficient of the polynomial.

7. $f(x) = 3x^5 - 3x^2 - 8$

8. $f(x) = -2x^7 - 3x$

9. $f(x) = 8x^4$

10. $f(x) = 24 - 4x + \frac{1}{3}x^2$

11. $f(x) = 3x + 5 - x^2\sqrt{5}$

12. $f(x) = -4x^5 - 4x + 7x^8 + 3x^9$

Write the function in standard form.

13. $f(x) = 3x^2 - 5 + 2x^3$

14. $f(x) = 3 + 2x - 5x^2$

15. $f(x) = 3x + 5x^2 - 3 + 2x^3$

16. $f(x) = 14 + 3x - 5x^2$

17. $f(x) = 6x - 5x^4 + 2$

18. $f(x) = x + 3 - 5x^3 + 7x^2$

Use direct substitution to evaluate the polynomial function for the given value of *x*.

19. $f(x) = 3 - x^2 + 4x - x^3,\ x = 2$

20. $f(x) = 3x^2 + 5x - 2x^5 + x^4,\ x = -1$

21. $f(x) = 7x + 2x^2 - 5,\ x = 3$

22. $f(x) = -x^2 + 5x + 22,\ x = -4$

Use what you know about end behavior to match the polynomial with its graph.

23. $f(x) = 2x^4 + 2x - 1$

24. $f(x) = -2x^3 + x^2 - 3x + 3$

25. $f(x) = -x^2 + 3x - 2$

26. $f(x) = 2x^3 + x^2 - 1$

A.

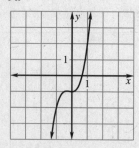

B.

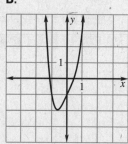

C.

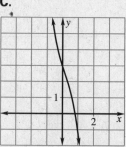

D.

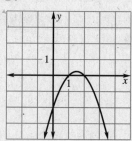

Computers   **In Exercises 27–29, use the following information.**

From 1990 to 1995, the number of computers per 1000 people in Germany can be modeled by $C = 75.32 + 14.93t + 0.99t^2$ where C is the number of computers per 1000 people and t is the number of years since 1990.

27. Write the model in standard form.

28. State the degree and leading coefficient of the model.

29. Estimate the number of computers per 1000 people in the year 2000.

Practice B

For use with pages 329–336

Decide whether the function is a polynomial function. If it is, write the function in standard form and state the degree and leading coefficient.

1. $f(x) = 3x^2 - 2x^3 + 4x$

2. $f(x) = 3x^{-3} + 2x + 1$

3. $f(x) = 4\sqrt{x} - 2x + 7x^3 - 1$

4. $f(x) = 2x^5 - 3 + 7x^2$

5. $f(x) = \frac{1}{3}x + \frac{2}{3} - \frac{1}{6}x^2$

6. $f(x) = -x + \sqrt{5}x^4 + 2x^2 - 7$

Use direct substitution to evaluate the polynomial function for the given value of x.

7. $f(x) = 3x + 2, \ x = -3$

8. $f(x) = 2x^3 - 3x^2 + 5x - 1, \ x = 1$

9. $f(x) = 4x^2 - 5x + 2, \ x = 3$

10. $f(x) = -3x^4 + 2x^2 - 3x + 4, \ x = -2$

11. $f(x) = 3x^7 + 2x^6 - 5x + 8, \ x = 0$

12. $f(x) = 6x^3 - 2x^2 + 5x + 2, \ x = 2$

13. $f(x) = -2x^5 + 3x^3 - 2x + 5, \ x = -1$

14. $f(x) = -x^4 - 2x^3 + 4x^2 + 6x - 3, \ x = 3$

Use synthetic substitution to evaluate the polynomial function for the given value of x.

15. $f(x) = 2x^3 - 3x^2 + 4x + 2, \ x = 4$

16. $f(x) = -2x^4 + 3x^3 - 5x^2 + 2x - 6, \ x = -2$

17. $f(x) = -5x^4 + 3x^2 + 2x - 5, \ x = 1$

18. $f(x) = x^6 + 3x + 4, \ x = 2$

19. $f(x) = 2x^2 - 4x + 7, \ x = -3$

20. $f(x) = x^4 + 3x^3 - 2x^2 + 8x, \ x = -4$

21. $f(x) = -4x^3 + 2x^2 + 6x, \ x = 3$

22. $f(x) = -3x^3 + 5x^2 + 6x - 8, \ x = -1$

Graph the polynomial function.

23. $f(x) = -x^3 + 2$

24. $f(x) = 2x^4 + 1$

25. $f(x) = 2x^3 + 1$

26. $f(x) = 3 - x^2$

27. $f(x) = x^3 + 2x - 3$

28. $f(x) = x^4 + 2x^3 - 5x + 1$

29. $f(x) = 1 - x^2 - x^3$

30. $f(x) = 2 + x^2 - x^4$

31. $f(x) = x^3 + x^2 - 2$

32. *Value of the Dollar* From 1988 to 1998 the value of a dollar in 1998 dollars can be modeled by $V = 0.002t^2 - 0.06t + 1.37$ where V is the value of the dollar and t is the number of years since 1988. What was the value of a dollar in 1996 in terms of 1998 dollars?

33. *Preakness Stakes* From 1990 to 1998, the money received by the winning horse can be modeled by $W = 6266.2t^3 - 79,306.8t^2 + 295,834.9t + 157,544.5$ where W is the winnings and t is the number of years since 1990. How much did Silver Charm win in 1997?

Use direct substitution to evaluate the polynomial function for the given value of x.

1. $f(x) = 3x^3 - 4x^2 + x - 7$, $x = 2$

2. $f(x) = \frac{1}{2}x^2 - x + 3$, $x = -4$

3. $f(x) = \frac{3}{2}x^3 - \frac{1}{4}x^2 + 3x - 1$, $x = 2$

4. $f(x) = 2x^4 - 3x^2 + 5$, $x = \frac{1}{2}$

5. $f(x) = -x^4 + 2x^2 + 5$, $x = -\sqrt{5}$

6. $f(x) = 2x^6 - x^4 + 5x + 1$, $x = \sqrt{2}$

Use synthetic substitution to evaluate the polynomial function for the given value of x.

7. $f(x) = -2x^5 + 3x^4 + x^3 - x^2 + 6x + 3$, $x = -1$

8. $f(x) = 3x^4 - 2x^2 + 5$, $x = 2$

9. $f(x) = \frac{2}{3}x^3 - 4x^2 + \frac{1}{2}x + 2$, $x = -2$

10. $f(x) = \frac{1}{5}x^2 + 3x - \frac{1}{2}$, $x = 3$

11. $f(x) = 4x^3 - 2x^2 + x - 3$, $x = -\frac{1}{2}$

12. $f(x) = -x^3 + 3x + 7$, $x = \frac{1}{3}$

Graph the function.

13. $f(x) = 4 - x^3$

14. $f(x) = 3x^2 - 5$

15. $f(x) = x^3 + 3x - 1$

16. $f(x) = -2x^4 + 3x - 1$

17. $f(x) = 2x^7 + 1$

18. $f(x) = \frac{2}{3} + x^2$

19. $f(x) = \frac{x^3}{4} + 2x + 1$

20. $f(x) = \sqrt{3}\,x^4 + 5$

21. $f(x) = 2x^3 - \sqrt{7}$

22. Critical Thinking Give an example of a polynomial function f such that $f(x) \to \infty$ as $x \to -\infty$ and $f(x) \to \infty$ as $x \to \infty$.

23. Critical Thinking Give an example of a polynomial function f such that $f(x) \to \infty$ as $x \to -\infty$ and $f(x) \to -\infty$ as $x \to \infty$.

First-time Brides **In Exercises 24 and 25, use the following information.**

The median age of a female when she gets married for the first time in the United States from 1890 to 1996 can be modeled by

$$A = 0.001t^2 - 0.098t + 22.763$$

where A is the age and t is the number of years since 1890.

24. What was the median age of first time brides in 1950?

25. Describe the end behavior of the graph. From the end behavior, would you expect first time brides in 2000 to be older or younger than the brides in 1996?

Algebra 2
Chapter 6 Resource Book

NAME _____ DATE _____

Reteaching with Practice

For use with pages 329–336

GOAL **How to evaluate and graph a polynomial function**

> ### VOCABULARY
>
> A **polynomial function** has the form $f(x) = a_n x^n + a_{n-1} x^{n-1} + \cdots + a_1 x + a_0$, where $a_n \neq 0$ and the exponents are all whole numbers. For this polynomial function a_n is the **leading coefficient**, a_0 is the **constant term**, and n is the **degree.**
>
> A polynomial function is in **standard form** if its terms are written in descending order of exponents from left to right.
>
> **Synthetic substitution** is another method of evaluating a polynomial function, which is equivalent to evaluating the polynomial in nested form.
>
> The **end behavior** of a polynomial function's graph is the behavior of the graph as x approaches positive infinity or negative infinity.

EXAMPLE 1 *Identifying Polynomial Functions*

Decide whether the function is a polynomial function. If it is, write the function in standard form and state its degree, type, and leading coefficient.

a. $f(x) = 3x^{1/2} - 2x^2 + 5$ **b.** $f(x) = 3$ **c.** $f(x) = \sqrt{5} - x$

SOLUTION

a. The function is not a polynomial function because the term $3x^{1/2}$ has an exponent that is not a whole number.

b. The function is a polynomial function. It is already in standard form. It has degree 0, so it is a constant function. The leading coefficient is 3.

c. The function is a polynomial function. Its standard form is $f(x) = -x + \sqrt{5}$. It has degree 1, so it is a linear function. The leading coefficient is -1.

Exercises for Example 1

Decide whether the function is a polynomial function. If it is, write the function in standard form and state the degree, type, and leading coefficient.

1. $f(x) = x + 5x^3$ **2.** $f(x) = 6 - 4x + \pi x^4$ **3.** $f(x) = 1 + 3x - \frac{1}{2}x^{-2}$

EXAMPLE 2 *Using Synthetic Substitution*

Use synthetic substitution to evaluate $f(x) = 4x^4 + 2x^3 - x + 7$ when $x = -2$.

NAME _____ DATE _____

Reteaching with Practice

For use with pages 329–336

SOLUTION

Begin by writing the polynomial in standard form, inserting terms with coefficients of 0 for missing terms. Then write the coefficients of $f(x)$ in a row. Bring down the leading coefficients and multiply by -2. Write the result in the next column. Add the numbers in the column and bring down the result. Continue until you reach the end of the row.

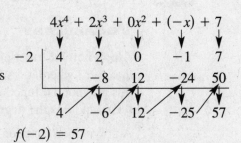

$$4x^4 + 2x^3 + 0x^2 + (-x) + 7$$

$$f(-2) = 57$$

Exercises for Example 2

Use synthetic substitution to evaluate the polynomial function for the given value of x.

4. $f(x) = x^3 + 5x^2 + 4x + 6, x = 2$ **5.** $f(x) = 2x^3 + x^4 + 5x^2 - x, x = -3$

6. $f(x) = x^3 - x^5 + 3, x = -1$ **7.** $f(x) = 5x^3 - 4x^2 - 2, x = 0$

EXAMPLE 3   *Graphing Polynomial Functions*

Graph (a) $f(x) = x^5 + 2x^2 - x + 4$ (b) $f(x) = -x^3 + 3x^2 + 6x - 2$

SOLUTION

Begin by making a table of values, including positive, negative, and zero values for x. Plot the points and connect them with a smooth curve. Then check the end behavior.

a.

x	-3	-2	-1	0	1	2	3
$f(x)$	-218	-18	6	4	6	42	262

The degree is odd and the leading coefficient is positive, so $f(x) \rightarrow -\infty$ as $x \rightarrow -\infty$ and $f(x) \rightarrow +\infty$ as $x \rightarrow +\infty$.

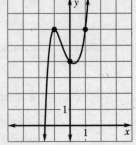

b.

x	-3	-2	-1	0	1	2	3
$f(x)$	34	6	-4	-2	6	14	16

The degree is odd and the leading coefficient is negative, so $f(x) \rightarrow +\infty$ as $x \rightarrow -\infty$ and

$f(x) \rightarrow -\infty$ as $x \rightarrow +\infty$.

Exercises for Example 3

Graph the polynomial function.

8. $f(x) = -x^5$ **9.** $f(x) = -x^4 + 1$ **10.** $f(x) = x^3 - 5$

Quick Catch-Up for Absent Students

For use with pages 329–337

The items checked below were covered in class on (date missed) ——————————

Lesson 6.2: Evaluating and Graphing Polynomial Functions

———— **Goal 1:** Evaluate a polynomial function. (pp. 329, 330)

Material Covered:

———— Example 1: Identifying Polynomial Functions

———— Student Help: Study Tip

———— Example 2: Using Synthetic Substitution

———— Example 3: Evaluating a Polynomial Function in Real-Life

Vocabulary:

polynomial function, p. 329   leading coefficient, p. 329

constant term, p. 329   degree, p. 329

standard form, p. 329   synthetic substitution, p. 330

———— **Goal 2:** Graph a polynomial function. (pp. 331, 332)

Material Covered:

———— Student Help: Look Back

———— Activity: Investigating End Behavior

———— Example 4: Graphing Polynomial Functions

———— Example 5: Graphing a Polynomial Model

Vocabulary:

end behavior, p. 331

Activity 6.2: Setting a Good Viewing Window (p. 337)

———— **Goal 1:** Use what you know about end behavior to choose an appropriate viewing window.

———— Student Help: Keystroke Help

———— Other (specify) ——

——

Homework and Additional Learning Support

———— Textbook (specify) _pp. 333–336_ ————————————————————————

——

———— Internet: Extra Examples at www.mcdougallittell.com

———— *Reteaching with Practice* worksheet (specify exercises)——————————————

———— *Personal Student Tutor* for Lesson 6.2

Lesson 6.2

NAME _____ DATE _____

Interdisciplinary Application

For use with pages 329–336

Banking

BUSINESS   Banks are an important part of daily business activity. They provide services such as safeguarding deposits, loans, automated teller machines (ATMs), checking accounts, and other financial services. Companies and individuals can borrow money from banks for purchases.

The federal government established the First Bank of the United States in 1791 largely in part to the debt left by the Revolutionary War. The First Bank proved to be somewhat profitable, earning income through its loans to government and private businesses at the time. But for most of the 1800s, the United States struggled to establish its own financial system, with various types of currency in existence.

The Federal Reserve Act of 1913 empowered the federal government to control bank reserves and the money supply. The Federal Reserve System serves as the central bank of the United States with most of the commercial banks belonging to it. In 1933, the Glass-Steagall Banking Act created the Federal Deposit Insurance Corporation (FDIC), which insures the deposits made into almost all commercial banks. Commercial banks are the most numerous type of bank in the United States with over 9000 in operation and over five trillion dollars in total assets.

In Exercises 1–3, use the following information.

The total assets A_1 (in billions of dollars) of insured commercial banks in the United States from 1985 through 1997 can be modeled by the equation $A_1 = 2.007t^3 - 54.09t^2 + 580.0t + 943$ where t is the number of years since 1980.

1. Sketch the graph of the polynomial function for the given years.

2. Describe the end behavior of the graph of the function. From the end behavior, would you expect the amount of total assets in 2010 to be more than or less than the amount in 1997?

3. Estimate the amount of total assets in 1995.

In Exercises 4–7, use the following information.

Data can sometimes be modeled by more than one polynomial. In this case, the total assets of insured commercial banks in the United States from 1985 through 1997 can be modeled by another equation, $A_2 = -0.0906t^4 + 5.996t^3 - 116.70t^2 + 992.2t - 10$ where t is the number of years since 1980.

4. Sketch the graph of the polynomial function A_2.

5. Using A_2, estimate the amount of total assets in 1995. How does your answer compare with the answer from Exercise 3?

6. Predict the future amount of total assets in 2010 using each model. What do you notice about your results?

7. Which model better represents the amount of total assets from 1985 through 1997? Explain.

NAME _____ DATE _____

Challenge: Skills and Applications

For use with pages 329–336

Use synthetic substitution to evaluate each polynomial for the given complex value of x.

1. $x^3 - 5x^2 + 3x - 1; x = i$

2. $-x^3 + 2x^2 + 3x - 5; x = 1 + i$

3. $x^4 + x - 1; x = i$

4. $2x^4 + x^3 - 1; x = -i$

5. Suppose $f(x) = ax^n + bx^{n-1} + cx^{n-2} + \ldots$, and $g(x) = rx^n + sx^{n-1} + tx^{n-2} + \ldots$, where $r \neq 0$. In this problem you will prove that

$$\frac{f(x)}{g(x)} \text{ gets close to } \frac{a}{r} \text{ as } x \to \infty.$$

a. Write the fraction $\dfrac{f(x)}{g(x)}$ with each term of the numerator and denominator divided by

x^n. How do you know that this new expression has the same values as the original fraction, provided $x \neq 0$?

b. Explain how you can easily tell what happens to the value of each term of the new expression as $x \to \infty$. Does the same argument work if you want to prove the same assertion with "$x \to \infty$" replaced by "$x \to -\infty$"?

6. You probably know that (*) a positive integer n is evenly divisible by 9 if and only if the sum of its digits is evenly divisible by 9. To prove this, write the following:

$$n = a \cdot 10^k + b \cdot 10^{k-1} + \ldots + r \cdot 10 + s$$

a. What do the numbers $a, b, \ldots , r, s$ represent? Explain how you know that for any positive integer j, $10^j = 9v + 1$ for some positive integer v.

b. Use the result of part (a) to show that you can "cast out 9s;" that is, the remainder when n is divided by 9 will be the same as the remainder when $a + b + \ldots + r + s$ is divided by 9.

c. Explain how your answer to part (b) proves the assertion (*).

7. a. Show that when the polynomial $P(x) = ax^3 + bx^2 + cx + d$ is evaluated for $x = k$ using synthetic substitution, the result is $ak^3 + bk^2 + ck + d$.

b. Use synthetic substitution and $P(x)$ in part (a) to show that if $a, b, c,$ and d are *integers* and $P(2) = 0$, then d is an *even* integer.

TEACHER'S NAME _____ CLASS _____ ROOM _____ DATE _____

Lesson Plan

2-day lesson (See *Pacing the Chapter,* TE pages 320C–320D)   For use with pages 338–344

GOALS
1. Add, subtract, and multiply polynomials.
2. Use polynomial operations in real-life problems.

State/Local Objectives _____

_____

✓ Check the items you wish to use for this lesson.

STARTING OPTIONS
____ Homework Check: TE page 333; Answer Transparencies
____ Warm-Up or Daily Homework Quiz: TE pages 338 and 336, CRB page 36, or Transparencies

TEACHING OPTIONS
____ Lesson Opener (Activity): CRB page 37 or Transparencies
____ Graphing Calculator Activity with Keystrokes: CRB pages 38–39
____ Examples: Day 1: 1–6, SE pages 338–339; Day 2: 7–8, SE page 340
____ Extra Examples: Day 1: TE page 339 or Transp.; Day 2: TE page 340 or Transp.
____ Closure Question: TE page 340
____ Guided Practice: SE page 341 Day 1: Exs. 1–11; Day 2: Ex. 12

APPLY/HOMEWORK
Homework Assignment
____ Basic   Day 1: 14–40 even, 46–50 even, 54–58 even, 62, 64; Day 2: 23–63 odd, 65–66, 70–71, 73–87 odd; Quiz 1: 1–27
____ Average   Day 1: 14–40 even, 46–50 even, 54–64 even; Day 2: 21–63 odd, 65–68, 70–71, 73–87 odd; Quiz 1: 1–27
____ Advanced   Day 1: 14–64 even; Day 2: 21–63 odd, 65–72, 73–87 odd; Quiz 1: 1–27

Reteaching the Lesson
____ Practice Masters: CRB pages 40–42 (Level A, Level B, Level C)
____ Reteaching with Practice: CRB pages 43–44 or Practice Workbook with Examples
____ Personal Student Tutor

Extending the Lesson
____ Applications (Real-Life): CRB page 46
____ Challenge: SE page 343; CRB page 47 or Internet

ASSESSMENT OPTIONS
____ Checkpoint Exercises: Day 1: TE page 339 or Transp.; Day 2: TE page 340 or Transp.
____ Daily Homework Quiz (6.3): TE page 343, CRB page 51, or Transparencies
____ Standardized Test Practice: SE page 343; TE page 343; STP Workbook; Transparencies
____ Quiz (6.1–6.3): SE page 344; CRB page 48

Notes _____

_____

_____

TEACHER'S NAME _____ CLASS _____ ROOM _____ DATE _____

Lesson Plan for Block Scheduling

1-day lesson (See *Pacing the Chapter,* TE pages 320C–320D)   **For use with pages 338–344**

GOALS 1. **Add, subtract, and multiply polynomials.**
2. **Use polynomial operations in real-life problems.**

State/Local Objectives _____

_____

_____

✓ Check the items you wish to use for this lesson.

STARTING OPTIONS

____ Homework Check: TE page 333; Answer Transparencies
____ Warm-Up or Daily Homework Quiz: TE pages 338 and 336,
         CRB page 36, or Transparencies

TEACHING OPTIONS

____ Lesson Opener (Activity): CRB page 37 or Transparencies
____ Graphing Calculator Activity with Keystrokes: CRB pages 38–39
____ Examples: 1–8: SE pages 338–340
____ Extra Examples: TE pages 339–340 or Transparencies
____ Closure Question: TE page 340
____ Guided Practice Exercises: SE page 341

APPLY/HOMEWORK
Homework Assignment

____ Block Schedule: 14–64 even, 65–68, 70–71, 73–87 odd; Quiz 1: 1–27

Reteaching the Lesson

____ Practice Masters: CRB pages 40–42 (Level A, Level B, Level C)
____ Reteaching with Practice: CRB pages 43–44 or Practice Workbook with Examples
____ Personal Student Tutor

Extending the Lesson

____ Applications (Real Life): CRB page 46
____ Challenge: SE page 343; CRB page 47 or Internet

ASSESSMENT OPTIONS

____ Checkpoint Exercises: TE pages 339–340 or Transparencies
____ Daily Homework Quiz (6.3): TE page 343, CRB page 51, or Transparencies
____ Standardized Test Practice: SE page 343; TE page 343; STP Workbook; Transparencies
____ Quiz (6.1–6.3): SE page 344; CRB page 48

Notes _____

_____

_____

CHAPTER PACING GUIDE	
Day	**Lesson**
1	6.1 (all); 6.2(all)
2	**6.3 (all)**
3	6.4 (all)
4	6.5 (all); 6.6(all)
5	6.7 (all); 6.8 (all)
6	6.9(all); Review Ch. 6
7	Assess Ch. 6; 7.1 (all)

Lesson 6.3

NAME _____ DATE _____

WARM-UP EXERCISES

For use before Lesson 6.3, pages 338–344

Simplify.

1. $2(x - 5)$

2. $-4(x^2 - 5x + 1)$

3. $(3x)(2x)$

4. $a \cdot a^3 \cdot a^3$

5. $7m^2 - 12m^2$

DAILY HOMEWORK QUIZ

For use after Lesson 6.2, pages 329–337

1. Tell whether $f(x) = \dfrac{3x}{5} - x^3 + 4x^2 + 7$ is a polynomial

function. If it is, write it in standard form and state the degree,
type, and leading coefficient.

2. Use synthetic division to evaluate $f(x) = 2x^4 + x^3 - 3x^2 + 6$
for $x = -2$.

3. Use what you know about end behavior to
match the graph with one of the polynomial
functions.

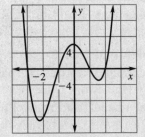

   (a) $f(x) = -2x^3 + 4x^2 + 6$

   (b) $f(x) = x^4 + x^3 - 7x^2 - x + 6$

   (c) $f(x) = -x^6 + 3x^5 - x^2 + 6$

   (d) $f(x) = x^5 - x^3 + 5x^2 + 6$

NAME _____ DATE _____

Activity Lesson Opener

For use with pages 338–344

SET UP: Work individually.

You can add or subtract polynomials by combining like terms, just as you have done with linear functions.

Garry Kasparov and a computer are both ___?___. To discover what two words go in this blank, match each sum or difference on the left with a simplified polynomial on the right. (A choice may be used more than once or not at all.)

____ **1.** $(x^2 + 3x + 9) + (x^2 - 3x + 6)$　　　　**A** $x^3 - x^2 - 3x$

____ **2.** $(-3x^2 + 6x - 7) + (3x^2 - x + 2)$　　　**B** $x^2 - 5x + 1$

____ **3.** $(3x^2 + 5x - 3) + (x^2 - 7)$　　　　　　**C** $2x^2 + 15$

____ **4.** $(x^2 + 3x - 4) + (2x + 5)$　　　　　　　**E** $4x^2 + 5x - 10$

____ **5.** $(x^2 + 15x) - (10x - 1)$　　　　　　　　**G** $x^3 - x^2 + 3x$

　　　　　　　　　　　　　　　　　　　　　　　　H $5x - 5$

____ **6.** $(2x^2 - 7x + 5) + (7x + 10)$　　　　　　**I** $2x^2 - 4x + 10$

____ **7.** $(x^2 + 8x - 3) - (x^2 + 3x + 2)$　　　　**L** $x^2 + 3x - 5$

____ **8.** $(2x^3 - x^2 + 4) - (x^3 + 3x + 4)$　　　**M** $-x^2 + 13x + 4$

____ **9.** $(8x + 7) - (x^2 - 5x + 3)$　　　　　　　**N** $x^3 - 2x^2 + 6x$

____ **10.** $(x^3 - 5x + 4) + (x^2 + 5x + 4)$　　　**O** $-x^2 - 2$

____ **11.** $(x^2 + x + 6) + (x^2 - 5x + 4)$　　　　**P** $x^3 + x^2 + 8$

____ **12.** $(3x - 7) - (x^2 + 3x - 5)$　　　　　　　**S** $x^2 + 5x + 1$

____ **13.** $(5x + 3) + (x^3 - 2x^2 + x - 3)$　　　**U** $x^2 + 5x - 1$

____ **14.** $(2x^2 + 5x - 6) - (x^2 - 7)$　　　　　　**Y** $x^3 - x^2 - 10x$

Graphing Calculator Activity

For use with pages 338–344

GOAL **To explore operations performed on polynomials**

To add or subtract polynomials, add or subtract the coefficients of the like terms.

$$(3x^2 + 5x - 4) + (2x^2 + 3x - 1) = (3 + 2)x^2 + (5 + 3)x + (-4 + -1)$$
$$= 5x^2 + 8x - 5$$

$$(3x^2 + 5x - 4) - (2x^2 + 3x - 1) = (3 - 2)x^2 + (5 - 3)x + (-4 - (-1))$$
$$= x^2 + 2x - 3$$

To multiply two polynomials, each term of the first polynomial must be multiplied by each term of the second polynomial.

$$(x - 3)(x + 1) = (x - 3)x + (x - 3)1 = x^2 - 3x + x - 3 = x^2 - 2x - 3$$

Activity

❶ Use a graphing calculator to verify that $(3x^2 + 5x - 4) + (2x^2 + 3x - 1) = 5x^2 + 8x - 5$.
Enter $(3x^2 + 5x - 4) + (2x^2 + 3x - 1)$ as Y1 and $5x^2 + 8x - 5$ as Y2. Graph equations Y1 and Y2 on the same screen. If the graphs coincide, then it is true that $(3x^2 + 5x - 4) + (2x^2 + 3x - 1) = 5x^2 + 8x - 5$.

❷ Use a graphing calculator to verify that $(3x^2 + 5x - 4) - (2x^2 + 3x - 1) = x^2 + 2x - 3$.
Enter $(3x^2 + 5x - 4) - (2x^2 + 3x - 1)$ as Y1 and $x^2 + 2x - 3$ as Y2. Graph equations Y1 and Y2 on the same screen. If the graphs coincide, then it is true that $(3x^2 + 5x - 4) - (2x^2 + 3x - 1) = x^2 + 2x - 3$.

❸ Use a graphing calculator to verify that $(x - 3)(x + 1) = x^2 - 2x - 3$.
Enter $(x - 3)(x + 1)$ as Y1 and $x^2 - 2x - 3$ as Y2. Graph equations Y1 and Y2 on the same screen. If the graphs coincide, then it is true that $(x - 3)(x + 1) = x^2 - 2x - 3$.

Exercises

1. Perform the indicated operation.

a. $(2x^2 + 2x) + (x^2 + 4x)$

b. $(3x^2 + 4x) - (2x^2 + 7x)$

c. $(5x^2 - 4x + 9) - (3x^2 - 8x - 3)$

d. $(4x^3 + 5x^2 + 7x - 1) - (2x^3 - 9x - 3)$

e. $(x - 2)(x + 5)$

f. $(2x - 5)(3x + 1)$

2. Use a graphing calculator to verify your answers from Exercise 1.

Algebra 2
Chapter 6 Resource Book

NAME _____   DATE _____

Graphing Calculator Activity

For use with pages 338–344

TI-82

Step 1:

Y= (3 X,T,θ x² + 5 X,T,θ −
4) + (2 X,T,θ x² + 3 X,T,θ
− 1) ENTER
5 X,T,θ x² + 8 X,T,θ − 5 ENTER
ZOOM 6

Step 2:   Y= CLEAR (3 X,T,θ x² +
5 X,T,θ − 4) − (2 X,T,θ x² +
3 X,T,θ − 1) ENTER CLEAR
X,T,θ x² + 2 X,T,θ − 3 ENTER
GRAPH

Step 3:   Y= CLEAR (X,T,θ − 3)
× (X,T,θ + 1) ENTER CLEAR
X,T,θ x² − 2 X,T,θ − 3 ENTER
GRAPH

TI-83

Step 1:

Y= (3 X,T,θ,n x² + 5 X,T,θ,n −
4) + (2 X,T,θ,n x² + 3 X,T,θ,n
− 1) ENTER
5 X,T,θ,n x² + 8 X,T,θ,n − 5 ENTER
ZOOM 6

Step 2:   Y= CLEAR (3 X,T,θ,n x² +
5 X,T,θ,n − 4) − (2 X,T,θ,n x²
+ 3 X,T,θ,n − 1) ENTER CLEAR
X,T,θ,n x² + 2 X,T,θ,n − 3 ENTER
GRAPH

Step 3:   Y= CLEAR (X,T,θ,n − 3)
× (X,T,θ,n + 1) ENTER CLEAR
X,T,θ,n x² − 2 X,T,θ,n − 3 ENTER
GRAPH

SHARP EL-9600c

Step 1:

Y= (3 X/θ/T/n x² + 5 X/θ/T/n −
4) + (2 X/θ/T/n x² + 3 X/θ/T/n
− 1) ENTER
5 X/θ/T/n x² + 8 X/θ/T/n − 5 ENTER
ZOOM [A] 5

Step 2:   Y= CL (3 X/θ/T/n x² +
5 X/θ/T/n − 4) − (2 X/θ/T/n x²
+ 3 X/θ/T/n − 1) ENTER CL
X/θ/T/n x² + 2 X/θ/T/n − 3 ENTER
GRAPH

Step 3:   Y= CL (X/θ/T/n − 3) ×
(X/θ/T/n + 1) ENTER CL
X/θ/T/n x² − 2 X/θ/T/n − 3 ENTER
GRAPH

CASIO CFX-9850GA PLUS

From the main menu, choose GRAPH.

Step 1:

(3 X,θ,T x² + 5 X,θ,T −
4) + (2 X,θ,T x² + 3 X,θ,T
− 1) EXE
5 X,θ,T x² + 8 X,θ,T − 5 EXE
SHIFT F3 F3 F6

Step 2:   (3 X,θ,T x² + 5 X,θ,T −
4) − (2 X,θ,T x² + 3 X,θ,T
− 1) EXE
X,θ,T x² + 2 X,θ,T − 3 EXE
F6

Step 3:   (X,θ,T − 3) × (
X,θ,T + 1) EXE
X,θ,T x² − 2 X,θ,T − 3 EXE
F6

Lesson 6.3

NAME _____ DATE _____

Practice A

For use with pages 338–344

Find the sum.

1. $\quad x^2 + 2x + 5$
$\quad + 2x^2 + 3x + 1$
_____

2. $\quad 6x^3 - 2x^2 - x + 3$
$\quad + -4x^3 - 3x^2 + 2x + 1$
_____

3. $\quad 3x^4 + x^3 + x^2 - x - 9$
$\quad + x^4 + 2x^3 - 3x^2 + 5x + 1$
_____

4. $(3x^2 + 2x - 5) + (x^2 + 3x + 5)$

5. $(2x^5 - 3x^4 + 2x^3 + x^2 - x - 8) + (3x^5 - 2x)$

6. $(x^2 + 7x + 1) + (-3x^2 - 10x + 7)$

7. $(-5x^3 + 2x^2 + x - 3) + (x^3 + 4x^2 - 3x + 1)$

Find the difference.

8. $\quad x^2 + 4x - 3$
$\quad - (x^2 + 3x + 1)$
_____

9. $\quad 3x^3 \qquad - 2x + 1$
$\quad - (x^3 + 2x^2 - 4x + 3)$
_____

10. $\quad 5x + 3$
$\quad - (6x + 1)$
_____

11. $(x + 7) - (2x + 4)$

12. $(x^2 + 3x - 1) - (2x^2 + x + 6)$

13. $(3x^2 + 2x - 1) - (x^2 - 5x + 2)$

14. $(2x^3 - 4x^2 + 3x - 7) - (3x^3 + x^2 - 5x - 2)$

15. $(4x^5 - 3x^2 + 8) - (2x^5 + 2x^2 - 1)$

16. $(7x^{12} + 3x^8 - 2x + 1) - (8x^{12} - 2x^8 + 3x + 5)$

Find the product.

17. $2x^2 + 4x + 1$
$\quad \times \qquad x - 4$
_____

18. $-x^2 + 3x + 7$
$\quad \times \qquad x + 1$
_____

19. $\quad 7x - 6$
$\quad \times 2x + 3$
_____

20. $(x + 4)(x - 3)$

21. $(x - 6)(x - 2)$

22. $(x + 3)(x + 2)$

23. $(x + 1)(2x + 3)$

24. $(2x - 1)(x - 5)$

25. $(3x + 2)(x - 1)$

26. $(x + 4)(x - 4)$

27. $(x - 7)(x + 7)$

28. $(x + 3)^2$

29. $(x + 6)^2$

30. $(x - 8)^2$

31. $(x - 4)^2$

Write the area of the figure as a polynomial in standard form.

32.
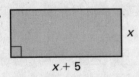
x
$x + 5$

33.
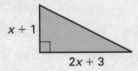
$x + 1$
$2x + 3$

34.
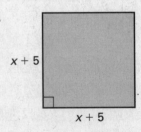
$x + 5$
$x + 5$

35. *Education*   For 1990 through 1996, the number of bachelor degrees D earned by people in the United States and the number of bachelor degrees W earned by women in the United States can be modeled by

$\quad D = 12829.86t + 1117893$

$\quad W = 9777.82t + 602005.12$

where t is the number of years since 1990. Find a model that represents the number of bachelor degrees M earned by men in the United States from 1990 through 1996.

NAME _____ DATE _____

Practice B

For use with pages 338–344

Find the sum or difference.

1. $(x^2 + 2x + 3) + (x^2 - 5)$

2. $(3x^3 - 2x^2 + x - 1) - (x^2 + 2x + 3)$

3. $(4x^2 + x - 3) - (2x^2 - 5x + 1)$

4. $(x^2 - 2x + 7) + (-5x^2 - 3)$

5. $(4x^3 - 2x) + (3x^3 - 3x^2 + 1)$

6. $(2x^3 - 3x^2 + x - 3) + (-x^2 + 2x + 4)$

7. $(2x^2 + 5 - x) - (4x^2 - 3x^4)$

8. $(1 - 3x + x^2 - x^3) + (3 + 2x^5 - 4x)$

9. $(4x^5 + 3x^4 - 5x + 1) - (2x^5 + x^2 - 3)$

10. $(6x^3 + 3x^2 - 5x - 1) - (7x^3 + 3x - 6)$

11. $(3x^3 - 2x^2 + 7x + 5) - (3x^3 - 2x^2 + 7x - 5)$

12. $(6x^2 + 3x - 7) + (2x^2 - 3x + 7)$

Find the product.

13. $x(3x - 1)$

14. $2x^2(x + 3)$

15. $x(3x^2 - x + 5)$

16. $(x + 5)(x - 2)$

17. $(x + 3)(x + 1)$

18. $(x - 4)(x - 1)$

19. $(2x + 1)(x + 5)$

20. $(3x + 1)(x - 4)$

21. $(2x - 3)(x - 1)$

22. $(2x + 5)(3x - 1)$

23. $(4x + 1)(2x - 3)$

24. $(5x - 4)(3x - 1)$

25. $(x + 1)(x^2 + x - 1)$

26. $(x - 3)(x^2 + 3x + 2)$

27. $(x + 2)(x^2 - 5x)$

28. $(x + 9)(x - 9)$

29. $(2x + 5)(2x - 5)$

30. $(x + 10)^2$

31. $(4x + 3)^2$

32. $(x - 12)^2$

33. $(3x - 8)^2$

34. **Floor Space** Find a polynomial that represents the total number of square feet for the floor plan shown below.

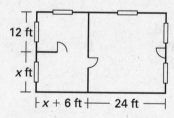

35. **Advertising** For 1980 through 1990, the amount of money A (in millions of dollars) spent on television and newspaper advertising can be modeled by

$$A = -16.2t^3 + 153t^2 + 3609.5t + 26,265.9$$

where t is the number of years since 1980. The amount of money n (in millions of dollars) spent on newspaper advertising can be modeled by

$$n = -30t^2 + 2257t + 14,761.8$$

where t is the number of years since 1980. Write a model that represents the amount of money v (in millions of dollars) spent on television advertising.

Practice C

For use with pages 338–344

Find the sum or difference.

1. $(2x^3 + 3x^2 - 5x + 2) + (-3x^3 - 6x^2 + 2)$ 2. $(4x^2 - 5x + 1) - (x^3 + x^2 - 3x + 4)$

3. $(-2x^2 + 7x - 7) - (3x^2 + 2x - 1)$ 4. $(2x^3 - 6x + 4) + (3x^4 - 2x^3 + 4x^2 + 1)$

5. $\left(\frac{1}{2}x^2 + 3x + 1\right) + \left(x^2 + \frac{2}{3}x - 3\right)$ 6. $\left(\frac{2}{5}x^2 + 2x - 1\right) + \left(\frac{3}{5}x^2 - 7x + \frac{1}{3}\right)$

7. $\left(\frac{1}{5}x^3 + 3x^2 + \frac{4}{3}\right) - \left(\frac{1}{2}x^3 + 2x + \frac{1}{3}\right)$ 8. $\left(\frac{3}{8}x^2 + \frac{2}{3}x - 5\right) - \left(\frac{3}{4}x^2 + \frac{1}{2}x\right)$

Find the product.

9. $(3x + 5)(2x + 5)$ 10. $(x - 7)(5x + 3)$ 11. $(3x - 4)(8x - 1)$

12. $(x + 1)(x^2 - 2x + 1)$ 13. $(x + 1)(2x^2 + 3x - 4)$ 14. $(2x - 1)(x^2 + 3x + 2)$

15. $(2x + 1)(x^2 - x - 3)$ 16. $(-x^2 + 3)(x^2 + 6x - 2)$

17. $(x^3 + x^2 + 3)(x^2 - 4x + 3)$ 18. $(x^3 - 2x + 1)(x^3 + x^2 - 5)$

19. $(2x^3 + x)(x^4 + 3x^3 - 2x^2 + 1)$ 20. $(6x + 5)(6x - 5)$

21. $\left(\frac{1}{2}x + 7\right)\left(\frac{1}{2}x - 7\right)$ 22. $\left(\frac{4}{3}x + 5\right)^2$ 23. $(5x - 2)^2$

24. $\left(\frac{1}{3}x - \frac{2}{3}\right)^2$ 25. $(x + 2)^3$ 26. $(x - 3)^3$

27. $(2x + 1)^3$ 28. $(3x - 5)^3$ 29. $(2x + 3y)^3$

30. $(4x - 3y)(4x + 3y)$ 31. $(6x + y)^2$ 32. $(x - 4y)^2$

Find the product of the binomials.

33. $(x + 3)(x + 2)(x - 1)$ 34. $(x - 5)(x + 3)(x - 2)$

35. $(2x + 1)(x + 3)(x + 1)$ 36. $(2x - 3)(2x - 5)(x - 1)$

37. **IRS Collection**   The principal source of collections by the IRS include
individual income and profit taxes, corporation income and profit taxes,
employment taxes, estate and gift taxes, and other taxes. From 1992
through 1996, the amount of taxes collected in each of these categories
can be modeled by

$T = 7,810,103.714t^2 + 61,813,629.34t + 1,116,758,213$ (Total collected)

$C = 18,508,265.4t + 116,419,459.8$ (Corporate income and profit)

$E = 23,846,333.7t + 394,945,983.6$ (Employment)

$G = 133,820.25t^3 - 881,998.57t^2 + 2,915,045.54t + 11,328,112.36$ (Estate and gift tax)

$O = -948,356.5t^3 + 4,663,117.93t^2 - 1,314,761.71t + 33,364,964.86$ (Other taxes)

where T, C, E, G and O are in thousands of dollars and t is the number of years since 1992.

Write a model that represents the individual income and profit taxes I (in thousands of dollars)
from 1992 to 1996.

NAME _____ DATE _____

Reteaching with Practice

For use with pages 338–344

GOAL Add, subtract, and multiply polynomials

VOCABULARY

Special Product Patterns

Sum and Difference   **Example**

$(a + b)(a - b) = a^2 - b^2$ $(x + 3)(x - 3) = x^2 - 9$

Square of a Binomial

$(a + b)^2 = a^2 + 2ab + b^2$ $(y + 4)^2 = y^2 + 8y + 16$

$(a - b)^2 = a^2 - 2ab + b^2$ $(3t^2 - 2)^2 = 9t^4 - 12t^2 + 4$

Cube of a Binomial

$(a + b)^3 = a^3 + 3a^2b + 3ab^2 + b^3$ $(x + 1)^3 = x^3 + 3x^2 + 3x + 1$

$(a - b)^3 = a^3 - 3a^2b + 3ab^2 - b^3$ $(p - 2)^3 = p^3 - 6p^2 + 12p - 8$

EXAMPLE 1 *Adding Polynomials Horizontally and Vertically*

Add the polynomials.

a. $(2x^2 + 8x + 4) + (x^2 - 8x - 2) = 2x^2 + x^2 + 8x - 8x + 4 - 2$

$$= 3x^2 + 2$$

b. $\quad 2x^3 - 3x^2 + x - 3$

$\quad + \quad\quad -x^2 + 2x + 4$

$\quad\overline{\quad 2x^3 - 4x^2 + 3x + 1}$

Exercises for Example 1

Find the sum.

1. $(5x^2 + 2x + 1) + (4x^2 + 3x - 8)$ **2.** $(9x^3 - 5) + (-11x^3 + 4)$

3. $(4x^2 + x + 6) + (7x - 10)$ **4.** $(14 - 6x) + (8x - 5)$

EXAMPLE 2 *Subtracting Polynomials Horizontally and Vertically*

a. $(3x^3 - 2x^2 + x) - (x^2 + 2x - 3) = 3x^3 - 2x^2 + x - x^2 - 2x + 3$ Add the opposite.

$$= 3x^3 - 3x^2 - x + 3$$

b. $\quad\quad 4x^2 + x - 3 \quad\quad\longrightarrow\quad\quad 4x^2 + x - 3$ Add the opposite.

$\quad - (2x^2 - 5x + 1) \quad\quad\quad\quad\quad \underline{-\ 2x^2 + 5x - 1}$

$\quad\quad\quad\quad\quad\quad\quad\quad\quad\quad\quad\quad\quad 2x^2 + 6x - 4$

LESSON
6.3
CONTINUED

NAME _____ DATE _____

Reteaching with Practice
For use with pages 338–344

Exercises for Example 2

Find the difference.

5. $(5x^2 - 6x - 1) - (4x^2 - 2x + 1)$ **6.** $(5x^3 + 7x + 8) - (x^3 - 6x + 4)$

7. $(-8x^2 + x + 5) - (2x^2 - 3)$ **8.** $(x^2 - 2x + 7) - (-5x^2 - 3)$

EXAMPLE 3 *Multiplying Polynomials Horizontally and Vertically*

Find the product of the polynomials.

a. $(x - 3)(x^2 + 3x + 2) = (x - 3)x^2 + (x - 3)3x + (x - 3)2$

$$= x^3 - 3x^2 + 3x^2 - 9x + 2x - 6$$

$$= x^3 - 7x - 6$$

b.
$$3x^2 - 2x + 1$$

$\underline{\times \qquad\qquad x + 2}$

$6x^2 - 4x + 2$ Multiply $3x^2 - 2x + 1$ by 2.

$\underline{3x^3 - 2x^2 + \ x}$ Multiply $3x^2 - 2x + 1$ by x.

$3x^3 + 4x^2 - 3x + 2$ Combine like terms.

Exercises for Example 3

Find the product of the polynomials.

9. $3x(x^2 + x - 2)$ **10.** $-2x(1 - x - x^2)$

11. $(x - 2)(x^2 + 2x + 4)$ **12.** $(2x + 1)(x^2 - x - 3)$

EXAMPLE 4 *Using Special Product Patterns*

Multiply the polynomials.

a. $(5x + 3)(5x - 3) = (5x)^2 - 3^2$ Sum and difference

$$= 25x^2 - 9$$

b. $(2x + y)^2 = (2x)^2 + 2(2x)(y) + (y)^2$ Square of a binomial

$$= 4x^2 + 4xy + y^2$$

c. $(2x - 1)^3 = (2x)^3 - 3(2x)^2(1) + 3(2x)(1)^2 - (1)^3$ Cube of a binomial

$$= 8x^3 - 12x^2 + 6x - 1$$

Exercises for Example 4

Find the product.

13. $(x + 5)(x - 5)$ **14.** $(2x + 7)(2x - 7)$ **15.** $(x + 6)^2$

16. $(x - 3)^2$ **17.** $(x - 1)^3$ **18.** $(x + 2)^3$

NAME _____ DATE _____

Quick Catch-Up for Absent Students

For use with pages 338–344

The items checked below were covered in class on (date missed) _____

Lesson 6.3: Adding, Subtracting, and Multiplying Polynomials

____ **Goal 1:** Add, subtract, and multiply polynomials. (pp. 338, 339)

Material Covered:

____ Example 1: Adding Polynomials Vertically and Horizontally

____ Example 2: Subtracting Polynomials Vertically and Horizontally

____ Student Help: Look back

____ Example 3: Multiplying Polynomials Vertically

____ Example 4: Multiplying Polynomials Horizontally

____ Student Help: Look Back

____ Example 5: Multiplying Three Binomials

____ Example 6: Using Special Product Patterns

____ **Goal 2:** Use polynomial operations in real-life problems. (p. 340)

Material Covered:

____ Example 7: Subtracting Polynomial Models

____ Example 8: Multiplying Polynomial Models

____ Other (specify) _____

_____

Homework and Additional Learning Support

____ Textbook (specify) _pp. 341–344_____

_____

____ Internet: Extra Examples at www.mcdougallittell.com

____ *Reteaching with Practice* worksheet (specify exercises)_____

____ *Personal Student Tutor* for Lesson 6.3

Real-Life Application:
When Will I Ever Use This?

For use with pages 338–344

Constructing a Pyramid

For a class project, you and your classmates are constructing two pyramids using aluminum cans. One of your pyramids will have a triangular base and the other will have a square base as shown below.

 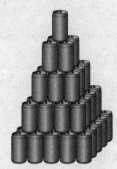

The formula for the total number of cans C in a triangular-base pyramid with n layers is $C = \frac{1}{6}n^3 + \frac{1}{2}n^2 + \frac{1}{3}n$. For a square-base pyramid, it is $C = \frac{1}{3}n^3 + \frac{1}{2}n^2 + \frac{1}{6}$.

1. Complete the table shown below for each type of pyramid. Round your answers to the nearest integer if necessary.

Number of layers, n	2	3	4	5	6	7	8	9	10	11	12
Total number of cans in pyramid, C											

2. Sketch the graphs of $C = \frac{1}{6}n^3 + \frac{1}{2}n^2 + \frac{1}{3}n$ and $C = \frac{1}{3}n^3 + \frac{1}{2}n^2 + \frac{1}{6}$ for $1 \le n \le 12$. How do the graphs compare?

3. Suppose you are constructing a triangular-base pyramid and you already have n layers stacked. In terms of n, how many total cans will be in the pyramid if another layer is added? Write your result in standard form.

4. Using the expression found in Exercise 3, explain how to find the number of cans in the added layer in terms of n. Write your result in standard form.

5. Redo Exercise 3 for a square-base pyramid.

6. Using the expression found in Exercise 5, explain how to find the number of cans in the added layer of a square-base pyramid in terms of n. Write your result in standard form.

7. How many more cans would be in a 20-layer square-base pyramid than a 20-layer triangular-base pyramid?

8. How many more cans are in the sixteenth layer of a square-base pyramid than the eighteenth layer of a triangular-base pyramid?

Challenge: Skills and Applications

For use with pages 338–344

Simplify.

1. $(p + q)(p - q)(p^2 + q^2)$

2. $(x - y)(x + y)(x^4 + x^2y^2 + y^4)$

3. $(a + b - c)(a + b + c)$

4. $(u^2 + v - w)(u^2 - v + w)$

5. a. Simplify the product $(x + y)(x^2 - xy + y^2)$.

b. Simplify the product $(x + y)(x^4 - x^3y + x^2y^2 - xy^3 + y^4)$.

c. Based on your answers to parts (a) and (b), write a general formula. Use "$2n$" to represent a general even integer and "$2n + 1$" to represent a general odd integer, and use ". . ." for missing terms.

6. a. Expand $(a + b)^4$.

b. Use the formula you wrote in part (a) to simplify $(x^2 + 2)^4$.

c. Multiply the expanded form you found in part (a) by $(a + b)$ to get a formula for $(a + b)^5$. When combining like terms, express each coefficient (except the first and last) *as a sum of 2 coefficients* from the expansion of $(a + b)^4$. (For example, write "$(1 + 2)$" instead of "3").

d. Based on the way you wrote $(a + b)^5$ in part (c), state a rule for finding the coefficients of $(a + b)^n$, given the coefficients of $(a + b)^{n-1}$.

7. Explain how you know that $(n + 1)^k - n^k$ has degree $k - 1$.

8. In this problem, you will find a formula for $f(n) = 0^2 + 1^2 + 2^2 + 3^2 + \cdots + n^2$.

a. Assume that $f(n)$ can be expressed as a polynomial in the variable n of degree 3: $f(n) = an^3 + bn^2 + cn + d$. Use $f(0)$ to explain why $d = 0$.

b. Explain why $f(n + 1) - f(n) = (n + 1)^2$.

c. Expand $f(n + 1) - f(n)$ as given in part (a).

d. Use the equation in part (b) to find a, b, and c by equating the coefficients of like powers of n on both sides. Write the formula you have discovered for $f(n)$.

NAME _____ DATE _____

Quiz 1

For use after Lessons 6.1–6.3

Evaluate the expression. (*Lesson 6.1*)

1. $\left(\frac{2}{3}\right)^{-2}$

2. $5^0 \cdot 4^{-2}$

3. $4^2 \cdot (4^3 \cdot 4^{-5})^2$

4. $\dfrac{(2^4 \cdot 4^3)^{-1}}{2^{-3} \cdot 4^{-4}}$

Simplify the expression. (*Lesson 6.1*)

5. $(4x^3y^5)^{-3}$

6. $(x^4y^{-3})(x^3y^{-2})^2$

7. $\left(\dfrac{2x^2}{y^3}\right)^{-2}$

8. $\dfrac{x^7y^{-4}}{x^{-2}y^{-6}}$

Graph the polynomial function. (*Lesson 6.2*)

9. $f(x) = x^4 + 3$

10. $f(x) = 2x^3 + 4x - 1$

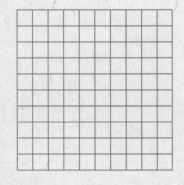
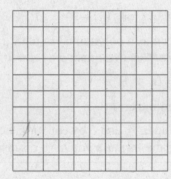

Perform the indicated operation. (*Lesson 6.3*)

11. $(-6x^2 + 5x - 2) - (7x^2 - 5x - 8)$

12. $(x + 3)(3x^2 + 2x - 1)$

13. $(x - 4)^3$

14. $(3x^2 + 2)^2$

Answers

1. _____
2. _____
3. _____
4. _____
5. _____
6. _____
7. _____
8. _____
9. _Use grid at left._
10. _Use grid at left._
11. _____
12. _____
13. _____
14. _____

Lesson 6.3

Lesson Plan

2-day lesson (See *Pacing the Chapter,* TE pages 320C–320D)   For use with pages 345–351

GOALS 1. **Factor polynomial expressions.**
2. **Use factoring to solve polynomial equations.**

State/Local Objectives _____

_____

✓ **Check the items you wish to use for this lesson.**

STARTING OPTIONS

_____ Homework Check: TE page 341; Answer Transparencies
_____ Warm-Up or Daily Homework Quiz: TE pages 345 and 343, CRB page 51, or Transparencies

TEACHING OPTIONS

_____ Motivating the Lesson: TE page 346
_____ Lesson Opener (Application): CRB page 52 or Transparencies
_____ Examples: Day 1: 1–3, SE page 346; Day 2: 4–5, SE page 347
_____ Extra Examples: Day 1: TE page 346 or Transp.; Day 2: TE page 347 or Transp.
_____ Closure Question: TE page 347
_____ Guided Practice: SE page 348 Day 1: Exs. 1–2, 4–10; Day 2: Exs. 3, 11–17

APPLY/HOMEWORK

Homework Assignment

_____ Basic   Day 1: 18–22 even, 24–27, 28–40 even, 46–68 even; Day 2: 29–59 odd, 63–83 odd, 88–90, 95–99 odd
_____ Average   Day 1: 18–27, 28–68 even; Day 2: 29–61 odd, 75–85 odd, 88–90, 95–99 odd
_____ Advanced   Day 1: 18–27, 28–74 even, 81–83; Day 2: 29–79 odd, 84–93, 95–99 odd

Reteaching the Lesson

_____ Practice Masters: CRB pages 53–55 (Level A, Level B, Level C)
_____ Reteaching with Practice: CRB pages 56–57 or Practice Workbook with Examples
_____ Personal Student Tutor

Extending the Lesson

_____ Cooperative Learning Activity: CRB page 59
_____ Applications (Interdisciplinary): CRB page 60
_____ Math & History: SE page 351; CRB page 61; Internet
_____ Challenge: SE page 350; CRB page 62 or Internet

ASSESSMENT OPTIONS

_____ Checkpoint Exercises: Day 1: TE page 346 or Transp.; Day 2: TE page 347 or Transp.
_____ Daily Homework Quiz (6.4): TE page 350, CRB page 65, or Transparencies
_____ Standardized Test Practice: SE page 350; TE page 350; STP Workbook; Transparencies

Notes _____

_____

_____

TEACHER'S NAME _____ CLASS _____ ROOM _____ DATE _____

Lesson Plan for Block Scheduling

1-day lesson (See *Pacing the Chapter,* TE pages 320C–320D)   For use with pages 345–351

GOALS   1. **Factor polynomial expressions.**
2. **Use factoring to solve polynomial equations.**

State/Local Objectives _____

_____

_____

✓ **Check the items you wish to use for this lesson.**

STARTING OPTIONS

_____ Homework Check: TE page 341; Answer Transparencies
_____ Warm-Up or Daily Homework Quiz: TE pages 345 and 343,
        CRB page 51, or Transparencies

TEACHING OPTIONS

_____ Motivating the Lesson: TE page 346
_____ Lesson Opener (Application): CRB page 52 or Transparencies
_____ Examples: 1–5: SE pages 346–347
_____ Extra Examples: TE pages 346–347 or Transparencies
_____ Closure Question: TE page 347
_____ Guided Practice Exercises: SE page 348

APPLY/HOMEWORK
Homework Assignment

_____ Block Schedule: 18–27, 28–68 even, 67–85 odd, 88–90, 95–99 odd

Reteaching the Lesson

_____ Practice Masters: CRB pages 53–55 (Level A, Level B, Level C)
_____ Reteaching with Practice: CRB pages 56–57 or Practice Workbook with Examples
_____ Personal Student Tutor

Extending the Lesson

_____ Cooperative Learning Activity: CRB page 59
_____ Applications (Interdisciplinary): CRB page 60
_____ Math & History: SE page 351; CRB page 61; Internet
_____ Challenge: SE page 350; CRB page 62 or Internet

ASSESSMENT OPTIONS

_____ Checkpoint Exercises: TE pages 346–347 or Transparencies
_____ Daily Homework Quiz (6.4): TE page 350, CRB page 65, or Transparencies
_____ Standardized Test Practice: SE page 350; TE page 350; STP Workbook; Transparencies

CHAPTER PACING GUIDE	
Day	**Lesson**
1	6.1 (all); 6.2(all)
2	6.3 (all)
3	**6.4 (all)**
4	6.5 (all); 6.6(all)
5	6.7 (all); 6.8 (all)
6	6.9(all); Review Ch. 6
7	Assess Ch. 6; 7.1 (all)

Notes _____

_____

_____

WARM-UP EXERCISES

For use before Lesson 6.4, pages 345–351

Factor.

1. $4x^2 - 24x$

2. $2x^2 + 11x - 21$

3. $4x^2 - 36x + 81$

Solve.

4. $x^2 + 10x + 25 = 0$

5. $6x^2 + x = 15$

DAILY HOMEWORK QUIZ

For use after Lesson 6.3, pages 338–344

Find the sum or difference.

1. $(9x^2 + 7x^3 - x^4 - 9) + (5x^4 - 11x^3 + 6x^2)$

2. $(6 - 4x^2 + 5x^3) - (7x^2 - x^4 - 12x^3)$

Find the product.

3. $(2x - 3)(2x - 5)$

4. $(x - 1)(-4x^3 + x^2 - 10)$

5. $(5x + 2y)^3$

6. $(x - 6)(2x + 1)(5x - 3)$

Application Lesson Opener

For use with pages 345–351

A small box measures $(x - 2)$ cm by x cm by $(x + 8)$ cm. The volume of the box is 96 cm². Follow the steps below to find the dimensions of the box.

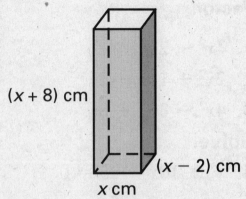

$(x + 8)$ cm

$(x - 2)$ cm

x cm

1. Write an equation to represent this situation.

2. One side of your equation is probably written in factored form. Explain why this does *not* help you to solve the equation.

3. Rewrite the equation in standard form. Your equation should be in the form $ax^3 + bx^2 + cx + d = 0$.

4. Factor $ax^3 + bx^2$, where a and b are the coefficients from your equation.

5. Factor $cx + d$, where c and d are the coefficients from your equation.

6. Compare your answers to Exercises 4 and 5. What similarity do you notice?

7. Use your answers to Exercises 4–6 to rewrite the equation from Exercise 3 in factored form. Then solve the equation. What are the three solutions?

8. Of the three solutions you wrote, which one makes sense in the context of the box problem?

9. What are the dimensions of the box?

Lesson 6.4

NAME _____ DATE _____

Practice A

For use with pages 345–351

Match the polynomial with its factorization.

1. $x^4 - 16$

2. $x^3 + 2x^2 + 6x + 12$

3. $x^3 - 3x^2 - 4x + 12$

4. $5x^3 + 5$

5. $x^3 + 27$

6. $8x^3 + 1$

7. $16x^4 - 81$

A. $(x + 3)(x^2 - 3x + 9)$

B. $(2x + 1)(4x^2 - 2x + 1)$

C. $(x + 2)(x^2 + 6)$

D. $(x - 2)(x + 2)(x^2 + 4)$

E. $(x - 3)(x - 2)(x + 2)$

F. $5(x + 1)(x^2 - x + 1)$

G. $(4x^2 + 9)(2x - 3)(2x + 3)$

Factor the sum or difference of cubes.

8. $x^3 + 1$

9. $x^3 + 27$

10. $x^3 + 125$

11. $x^3 - 1$

12. $x^3 - 8$

13. $x^3 - 64$

Factor the polynomial by grouping.

14. $x^3 + 3x^2 + 2x + 6$

15. $x^3 - x^2 + 4x - 4$

16. $x^3 + 5x^2 + x + 5$

17. $x^3 - 6x^2 + x - 6$

18. $x^3 + 4x^2 + 3x + 12$

19. $x^3 - 5x^2 + 2x - 10$

Find the real-number solutions of the equation.

20. $x^2 + 2x = 0$

21. $x^3 - 3x^2 = 0$

22. $x^2 + 3x - 4 = 0$

23. $x^2 + 5x + 6 = 0$

24. $x^2 - 49 = 0$

25. $x^2 - 100 = 0$

26. $x^3 + 2x^2 - x - 2 = 0$

27. $x^3 - x^2 - 4x + 4 = 0$

28. $x^3 + x^2 - 9x - 9 = 0$

Match the equations for volume with the appropriate solid.

29. $V = x^3 - 4x$

30. $V = x^3 - 4x^2 + 4x$

31. $V = x^4 - 16$

A.

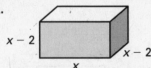

B.

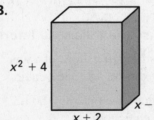

C.

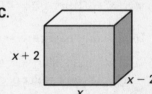

NAME _____ DATE _____

Practice B
For use with pages 345–351

Factor the sum or difference of cubes.

1. $x^3 + 64$

2. $x^3 + 216$

3. $x^3 - 1000$

4. $x^3 - 343$

5. $8x^3 - 1$

6. $8x^3 - 125$

7. $27x^3 - 8$

8. $8x^3 + 1000$

9. $27x^3 + 512$

10. $64x^3 + 27$

11. $1000x^3 - 1$

12. $125x^3 + 64$

Factor the polynomial by grouping.

13. $x^3 - 3x^2 + 5x - 15$

14. $x^3 - 4x^2 + 2x - 8$

15. $x^3 + 2x^2 + 7x + 14$

16. $3x^3 - 12x^2 + 2x - 8$

17. $5x^3 + 5x^2 + x + 1$

18. $2x^3 - 12x^2 + 5x - 30$

19. $x^3 - 2x^2 - 4x + 8$

20. $x^3 - 5x^2 - 9x + 45$

21. $x^3 + x^2 - 16x - 16$

22. $4x^3 - 4x^2 - 9x + 9$

23. $16x^3 - 48x^2 - x + 3$

24. $9x^3 + 18x^2 - 4x - 8$

Factor the polynomial.

25. $16x^4 - 81$

26. $x^4 - 9$

27. $x^4 + 5x^2 + 6$

28. $x^4 + x^2 - 6$

29. $x^4 + 5x^2 - 24$

30. $x^4 - 7x^2 + 10$

31. $2x^4 - 200x^2$

32. $8x^4 - 18x^2$

33. $27x^4 - 3x^2$

34. $3x^4 - 3$

35. $2x^4 + 16x^2 + 24$

36. $-x^4 - 10x^2 - 21$

Find the real-number solutions of the equation.

37. $x^3 + 6x^2 - 4x - 24 = 0$

38. $x^3 - 8 = 0$

39. $8x^3 + 27 = 0$

40. $3x^3 - x^2 + 3x - 1 = 0$

41. $x^3 + 7x^2 + 4x + 28 = 0$

42. $x^3 - 5x^2 - 9x + 45 = 0$

43. $x^4 - x^2 - 12 = 0$

44. $x^4 + 6x^2 + 5 = 0$

45. $x^4 - 10x^2 + 9 = 0$

46. $x^4 - 4x^2 - 5 = 0$

47. $x^4 - 10x^2 + 24 = 0$

48. $x^4 - 10x^2 + 16 = 0$

Aquarium **In Exercises 49–52, use the following information.**

The aquarium shown at the right holds 5610 gallons of water.
Each gallon of water occupies approximately 0.13368 cubic feet.

49. How many cubic feet of water does the aquarium hold?
    (Round the result to the nearest cubic foot.)

50. Use the result from Exercise 49 to write an equation that
    represents the volume of the aquarium.

51. Find all real solutions of the equation in Exercise 50.

52. What are the dimensions of the aquarium?

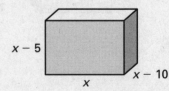

$x - 5$

$x - 10$

x

Practice C

For use with pages 345–351

Factor the polynomial.

1. $x^3 + 729$

2. $64x^3 - 125$

3. $2x^3 + 16$

4. $40x^3 + 5$

5. $16x^3 - 54$

6. $16x^4 - 81$

7. $x^4 - 64$

8. $3x^4 - 48$

9. $x^3 + 2x^2 - 49x - 98$

10. $x^3 - x^2 + 3x - 3$

11. $4x^3 - 24x^2 - x + 6$

12. $x^4 + 2x^3 - x - 2$

13. $8x^4 + 8x^3 - x - 1$

14. $x^4 + 5x^3 + x + 5$

15. $8x^4 + 8x^3 + 64x + 64$

16. $3x^3 + 12x^2 - 3x - 12$

17. $3x^4 + 6x^3 - 24x - 48$

18. $x^4 - 3x^3 - 4x^2 + 12x$

19. $x^6 + x^5 - x^4 - x^3$

20. $6x^4 + 12x^3 + 12x^2 + 24x$

21. $x^7 + x^6 + x^3 + x^2$

Find the real number solutions of the equation.

22. $x^3 - 1331 = 0$

23. $27x^3 + 8 = 0$

24. $4x^3 = 108$

25. $x^4 - 16 = 0$

26. $32x^4 = 2$

27. $x^3 + 12 = 3x^2 + 4x$

28. $x^3 + 8x^2 + 6x + 48 = 0$

29. $9x^3 + 27x^2 = 4x + 12$

30. $x^4 - 5x^3 = 8x - 40$

31. $27x^4 - 8 = 27x^3 - 8x$

32. $2x^3 - 6x^2 + 2x - 6 = 0$

33. $2x^4 - 10x^3 + 2x - 10 = 0$

34. $x^4 + 6x^3 = 9x^2 + 54x$

35. $x^6 + x^5 = x^4 + x^3$

36. $x^5 + 3x^4 = 8x^2 + 24x$

37. $5x^4 - 5x^3 - 20x^2 + 20x = 0$

38. $x^8 - 3x^7 = x^6 - 3x^5$

39. $x^3 = 2x^2 + 3x - 6$

40. $x^3 + 4x^2 = 7x + 28$

41. $2x^3 - 2x^2 = 5x - 5$

42. $x^4 - 36 = 0$

43. $x^5 - 100x = 0$

44. $3x^4 - 12 = 0$

45. $x^8 - 256 = 0$

46. *Manufacturing*   A tool shop is hired to make a metal mold in which plastic is injected to make a solid block. (See diagram below.) The finished plastic block should have a length that is 8 inches longer than the height. It should also have a width that is 2 inches shorter than the height. Each plastic block requires 96 cubic inches of plastic. If the sides of the mold are to be $\frac{1}{2}$ inch thick, how much metal is required to make the mold?

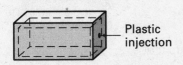

Plastic injection

Reteaching with Practice

For use with pages 345–351

GOAL How to factor polynomial expressions and use factoring to solve polynomial equations

---

VOCABULARY

Two special factoring patterns are given below.

Sum of Two Cubes   **Difference of Two Cubes**

$a^3 + b^3 = (a + b)(a^2 - ab + b^2)$ $a^3 - b^3 = (a - b)(a^2 + ab + b^2)$

To **factor by grouping,** group pairs of terms that have a common monomial factor together, factor out this common factor and look for a pattern. An expression of the form $au^2 + bu + c$ where u is any expression in x is said to be in **quadratic form.**

EXAMPLE 1 *Factoring the Sum or Difference of Cubes*

Factor each polynomial.

a. $64x^3 + 1 = (4x)^3 + 1^3$ Sum of two cubes

$ = (4x + 1)(16x^2 - 4x + 1)$

b. $54x^3 - 16 = 2(27x^3 - 8)$ Factor common monomial.

$ = 2[(3x)^3 - 2^3]$ Difference of two cubes

$ = 2(3x - 2)(9x^2 + 6x + 4)$

Exercises for Example 1

Factor the polynomial.

1. $x^3 + 125$ **2.** $x^3 - 343$ **3.** $64x^3 - 1$

4. $8x^3 + 27$ **5.** $3x^3 - 24$ **6.** $1000x^3 - 729$

EXAMPLE 2 *Factoring By Grouping*

Factor the polynomial $x^3 + x^2 - 4x - 4$.

SOLUTION

$x^3 + x^2 - 4x - 4 = (x^3 + x^2) - (4x + 4)$ Group terms.

$ = x^2(x + 1) - 4(x + 1)$ Factor out common monomial factor.

$ = (x^2 - 4)(x + 1)$ Factor out $(x + 1)$.

$ = (x + 2)(x - 2)(x + 1)$ Difference of two squares

Exercises for Example 2

Factor the polynomial by grouping.

7. $x^3 - x^2 - 9x + 9$ **8.** $x^3 - x + 5x^2 - 5$ **9.** $x^3 - 3x^2 - 16x + 48$

Reteaching with Practice

For use with pages 345–351

EXAMPLE 3 *Factoring Polynomials in Quadratic Form*

Factor each polynomial.

a. $36x^4 - 9x^2$　　　　　　　　　　**b.** $x^4 + 6x^2 + 9$

SOLUTION

a. $36x^4 - 9x^2 = 9x^2(4x^2 - 1)$　　　**b.** $x^4 + 6x^2 + 9 = (x^2)^2 + 2(x^2)(3) + 3^2$
$$= 9x^2[(2x)^2 - 1^2]$$　　　　　　　　　　$$= (x^2 + 3)^2$$
$$= 9x^2(2x + 1)(2x - 1)$$

Exercises for Example 3

Factor the polynomial.

10. $25x^4 - 9$　　　　　　　　　　**11.** $x^4 - x^2 + 6$

12. $x^4 - 16x^2 + 64$　　　　　　　**13.** $49x^4 - 4$

EXAMPLE 4 *Solving a Polynomial Equation*

Solve $3x^3 + 21x = 24x^2$.

SOLUTION

$$3x^3 + 21x = 24x^2$$　　　Write original equation.

$$3x^3 - 24x^2 + 21x = 0$$　　Rewrite in standard form.

$$3x(x^2 - 8x + 7) = 0$$　　Factor common monomial.

$$3x(x - 7)(x - 1) = 0$$　　Factor trinomial.

$$x = 0, x = 7, x = 1$$　　Set each factor equal to 0 and solve for x.

The solutions are 0, 1, and 7.

Exercises for Example 4

Find the real-number solutions of the equation.

14. $3x^2 = 9x$　　　　　　　　　　**15.** $x^2 = 2x + 15$

16. $4x^3 = 16x$　　　　　　　　　**17.** $x^2 + 2x = 24$

Lesson 6.4

Quick Catch-Up for Absent Students

For use with pages 345–351

The items checked below were covered in class on (date missed) _____

Lesson 6.4: Factoring and Solving Polynomial Equations

_____ **Goal 1:** Factor polynomial expressions. (pp. 345, 346)

Material Covered:

_____ Activity: The Difference of Two Cubes

_____ Example 1: Factoring the Sum or Difference of Cubes

_____ Example 2: Factoring by Grouping

_____ Example 3: Factoring Polynomials in Quadratic Form

Vocabulary:

factor by grouping, p. 346   quadratic form, p. 346

_____ **Goal 2:** Use factoring to solve polynomial equations. (p. 347)

Material Covered:

_____ Student Help: Study Tip

_____ Example 4: Solving a Polynomial Equation

_____ Example 5: Solving a Polynomial Equation in Real Life

_____ Other (specify) _____

_____

Homework and Additional Learning Support

_____ Textbook (specify) pp. 348–351 _____

_____

_____ Internet: Extra Examples at www.mcdougallittell.com

_____ *Reteaching with Practice* worksheet (specify exercises)_____

_____ *Personal Student Tutor* for Lesson 6.4

Cooperative Learning Activity

For use with pages 345–351

GOAL **To use the concept of solving polynomial equations**

Materials: paper, pencil, calculator

Background

You know how to solve first and second degree equations by factoring. Higher degree polynomial equations can also be solved by factoring.

Instructions

1 Solve: $x + 5 = -7$

2 Solve: $x^2 - 5x + 4 = 0$

3 Solve: $x^3 + 2x^2 - x = 2$

4 Solve: $2x^4 + x^3 - 8x^2 - x + 6 = 0$

5 Solve: $x^5 - x^4 - 7x^3 + 7x^2 + 12x - 12 = 0$

Analyzing the Results

1. Were you able to solve the polynomials in Steps 3, 4, and 5? How did your group solve them?

2. Was there any pattern that developed involving the degree of the equation and the number of solutions?

3. Write a sentence that describes the pattern.

NAME _____ DATE _____

Interdisciplinary Application

For use with pages 345–351

Greenhouses

BIOLOGY   A greenhouse is a building made of glass or polyethylene that can be used to grow flowers, shrubs, and other types of plants. Temperature, sunlight, moisture, and other conditions that are essential for plant growth can be regulated inside a greenhouse, which enables a person to grow plants year-round despite the harsh conditions of the climate.

The history of the greenhouse can be traced as far back as 30 A.D. when it is said that the Roman emperor Tiberius was instructed by his doctor to eat a cucumber a day. Tiberius's servants, in a sense, created their own greenhouses by growing cucumbers in large pots covered with sheets of mica. It was not until the sixteenth century that the first true greenhouses were built. European explorers would bring back rare tropical plants to Europe's colder climates, and in order to keep the plants alive, greenhouses (known then as botanical gardens) had to be constructed.

During this time, greenhouses played an integral part in the research and classification of plants. The science of botany, which is the branch of biology that studies the life, growth, and structure of plants, generated a tremendous amount of research into the cultivation of all the new plants being discovered. Universities as well as individuals from all over Europe began to collect and study all the varieties of plants.

In Exercises 1–4, use the following information.

The "lower portion" of the greenhouse shown
at the right has a volume of 528 cubic yards.
It has a height of x yards, a width of $5x - 4$
yards, and a length of $20x - 44$ yards.

1. Write an equation equal to zero that will
   allow you to solve for x.

2. What is x?

3. What are the dimensions of the greenhouse?

4. Suppose the "lower portion" of the greenhouse shown has a volume of 609
   cubic yards, and dimensions of x yards high, $12x - 29$ yards wide, and
   $12x - 7$ yards long. What would its dimensions be?

NAME _____ DATE _____

Math and History Application

For use with page 351

HISTORY   From the time when Babylonian mathematicians used tables of sums of powers to solve polynomial equations to the present, polynomials have been useful functions.

Any function that you need can be approximated by a polynomial. If your very mathematical model doesn't require high accuracy, then you can use quadratic or cubic polynomials. But, if NASA wants to guide its robot very precisely, it might use polynomials of degree 8 or 10.

An advantage of polynomials is that they are built from addition, subtraction, and multiplication. In fact, you can think of a polynomial as a set of instructions for starting with a number x and building a new number $f(x)$ using only these three operations. Since calculators and computers can add, subtract, and multiply very fast, polynomials are ideal for programmed calculation. Not only can NASA's engineers define polynomials that will control a robot, but NASA's programmers can also write fast-executing code to tell the robot's onboard computer how to evaluate these polynomials.

Programmers use a clever kind of *reexpression* similar to the equation manipulation described in the Math and History feature on page 351. Here's an example. Suppose your polynomial $f(x)$ is $x^3 - 4x^2 + 2x - 17$. On your calculator, using $+, -, \times$, and memory, you *could* evaluate $f(5)$ this way:

$$5 \times 5 \times 5 - 4 \times 5 \times 5 + 2 \times 5 - 17.$$

This takes five multiplications, three plus or minus operations, and some memory operations. But there's a faster way:

$$[(5 - 4) \times 5 + 2] \times 5 - 17.$$

If you evaluate this working from the inside out you still need three plus or minus operations, but only *two* multiplications. The savings are even greater for high-degree polynomials.

MATH   The following exercises illustrate the reexpression of polynomials for evaluation. The method illustrated above is called nesting. Here's one more example: In general form, the nested version of $x^3 - 4x^2 + 2x - 17$ is $[(x - 4)x + 2]x - 17$.

1. Write a nested expression for $x^2 + 5x - 3$.

2. a. Write a nested expression for $r(x) = x^4 - 7x^3 + x^2 - x + 4$.

   b. To evaluate r for a particular input x, how many multiplications will you do if you use your nested expression? How many additions or subtractions?

   c. If you evaluate $r(x)$ directly, how many multiplications will you do?

3. Write a nested expression for $4x^3 - 2x + 1$.

4. How many multiplications are required for a nested evaluation of $x^n + a_1x^{n-1} + \cdots + a_n$?

NAME _____ DATE _____

Challenge: Skills and Applications

For use with pages 345–351

Factor each expression into linear factors.

Example $x(x + 5) - 2(x + 5)^2$

Solution $x(x + 5) - 2(x + 5)^2 = (x + 5)[x - 2(x + 5)] = (x + 5)(x - 2x - 10)$
$$= (x + 5)(-x - 10)$$

1. $3(x - 2)^2 + (x - 2)^3$ **2.** $2x^2(x + 3)^3 - x(x + 3)^4$

3. a. Factor $a^6 + 1 = (a^2)^3 + 1$ as a sum of two cubes.

   b. Find a factorization of $2^6 + 1$ given by the formula you wrote in part (a). Do the same for $2^{12} + 1$.

   c. Show that a number of the form $2^n + 1$ ($n > 0$) cannot be prime if n is a multiple of 3 (i.e. if $n = 3k$ for some integer k).

4. a. Factor $x^6 - 64 = (x^2)^3 - 64$ as a difference of two cubes. Then factor one of the factors as a difference of squares.

   b. Factor $x^6 - 64 = (x^3)^2 - 64$ as a difference of two squares, *then* factor the two factors: one as a sum of cubes and one as a difference of cubes.

   c. Compare the factors you get from the factorizations in parts (a) and (b), and disregard the factors in each that are obviously equal. Write an equation that equates the factor(s) in each expression that are different.

   d. Generalize the equation you wrote in part (c) with an arbitrary integer a instead of 2 (i.e. write the corresponding factors in the factorization of $x^6 - a^6$).

5. a. Use the fact that $x^2 + 1 = (x + i)(x - i)$ to factor $x^4 + 16$ into two expressions with complex coefficients, each of degree 2.

   b. Factor $x^2 + p^2$, where p is a positive integer. Does this factorization hold if p and/or p^2 is complex?

   c. Show that if $w = \dfrac{\sqrt{2}}{2}p(1 + i)$, then $w^2 = p^2 i$.

   d. Use your answer to part (b), in conjunction with the fact stated in part (c), to write your factorization from part (a) as a product of 4 linear factors.

TEACHER'S NAME _____ CLASS _____ ROOM _____ DATE _____

Lesson Plan

1-day lesson (See *Pacing the Chapter*, TE pages 320C–320D)   **For use with pages 352–358**

GOALS   1. **Divide polynomials and relate the result to the remainder theorem and the factor theorem.**
2. **Use polynomial division in real-life problems.**

State/Local Objectives _____

_____

✓ Check the items you wish to use for this lesson.

STARTING OPTIONS

____ Homework Check: TE page 348; Answer Transparencies
____ Warm-Up or Daily Homework Quiz: TE pages 352 and 350, CRB page 65, or Transparencies

TEACHING OPTIONS

____ Motivating the Lesson: TE page 353
____ Lesson Opener (Activity): CRB page 66 or Transparencies
____ Examples 1–5: SE pages 352–355
____ Extra Examples: TE pages 353–355 or Transparencies; Internet
____ Closure Question: TE page 355
____ Guided Practice Exercises: SE page 356

APPLY/HOMEWORK
Homework Assignment

____ Basic   16–20 even, 28–34 even, 40–56 even, 59–63 odd, 64, 67–81 odd
____ Average   16–20 even, 28–58 even, 59–65 odd, 67–81 odd
____ Advanced   16–58 even, 59–65, 67–81 odd

Reteaching the Lesson

____ Practice Masters: CRB pages 67–69 (Level A, Level B, Level C)
____ Reteaching with Practice: CRB pages 70–71 or Practice Workbook with Examples
____ Personal Student Tutor

Extending the Lesson

____ Applications (Real-Life): CRB page 73
____ Challenge: SE page 358; CRB page 74 or Internet

ASSESSMENT OPTIONS

____ Checkpoint Exercises: TE pages 353–355 or Transparencies
____ Daily Homework Quiz (6.5): TE page 358, CRB page 77, or Transparencies
____ Standardized Test Practice: SE page 358; TE page 358; STP Workbook; Transparencies

Notes _____

_____

_____

TEACHER'S NAME _____ CLASS _____ ROOM _____ DATE _____

Lesson Plan for Block Scheduling

Half-day lesson (See *Pacing the Chapter,* TE pages 320C–320D)   For use with pages 352–358

GOALS   1. **Divide polynomials and relate the result to the remainder theorem and the factor theorem.**
2. **Use polynomial division in real-life problems.**

State/Local Objectives _____

_____

_____

✓ **Check the items you wish to use for this lesson.**

STARTING OPTIONS
____ Homework Check: TE page 348; Answer Transparencies
____ Warm-Up or Daily Homework Quiz: TE pages 352 and 350,
        CRB page 65, or Transparencies

CHAPTER PACING GUIDE	
Day	Lesson
1	6.1 (all); 6.2(all)
2	6.3 (all)
3	6.4 (all)
4	**6.5 (all)**; 6.6(all)
5	6.7 (all); 6.8 (all)
6	6.9(all); Review Ch. 6
7	Assess Ch. 6; 7.1 (all)

TEACHING OPTIONS
____ Motivating the Lesson: TE page 353
____ Lesson Opener (Activity): CRB page 66 or Transparencies
____ Examples 1–5: SE pages 352–355
____ Extra Examples: TE pages 353–355 or Transparencies; Internet
____ Closure Question: TE page 355
____ Guided Practice Exercises: SE page 356

APPLY/HOMEWORK
Homework Assignment (See also the assignment for Lesson 6.6.)
____ Block Schedule: 16–20 even, 28–58 even, 59–65 odd, 67–81 odd

Reteaching the Lesson
____ Practice Masters: CRB pages 67–69 (Level A, Level B, Level C)
____ Reteaching with Practice: CRB pages 70–71 or Practice Workbook with Examples
____ Personal Student Tutor

Extending the Lesson
____ Applications (Real Life): CRB page 73
____ Challenge: SE page 358; CRB page 74 or Internet

ASSESSMENT OPTIONS
____ Checkpoint Exercises: TE pages 353–355 or Transparencies
____ Daily Homework Quiz (6.5): TE page 358, CRB page 77, or Transparencies
____ Standardized Test Practice: SE page 358; TE page 358; STP Workbook; Transparencies

Notes _____

_____

_____

_____

LESSON

6.5

Available as
a transparency

Lesson 6.5

NAME _____ DATE _____

WARM-UP EXERCISES

For use before Lesson 6.5, pages 352–358

Simplify.

1. $\dfrac{3x^3}{x^2}$

2. $\dfrac{5x^2}{x^2}$

Find the missing factor.

3. $x^2 - 2x - 63 = (x - 9)(?)$

4. $2x^2 + 13x + 15 = (x + 5)(?)$

DAILY HOMEWORK QUIZ

For use after Lesson 6.4, pages 345–351

Find the greatest common factor of the terms in the polynomial.

1. $14x^2 + 35x^3 - 21x$

2. $60x^4 + 15x^3 - 30x^2$

Factor the polynomial.

3. $8x^3 + 125$

4. $5x^3 + 10x^2 - x - 2$

5. $200x^6 - 2x^4$

LESSON
6.5

NAME _____ DATE _____

Activity Lesson Opener
For use with pages 352–358

SET UP: Work individually.

In Lesson 6.5, you will learn how to perform polynomial long division, which is closely related to long division with numbers.

1. Fill in digits to complete the division problem below.

```
      2 □ □
  □ □ 3 1 0 5
      2 4
      □ □
      □ □
      □ □ □
          □ □
          □ □ □
```

2. In this division problem,

   a. What is the *dividend*?

   b. What is the *divisor*?

   c. What is the *quotient*?

   d. What is the *remainder*?

3. How would you use multiplication to check the result of this division problem?

4. How did you decided what number should be subtracted at each step?

5. Would the results be different if you changed 3105 to 315? Explain the importance of using zero as a placeholder.

LESSON 6.5

Practice A

For use with pages 352–358

Write the polynomial form of the dividend, divisor, quotient, and remainder represented by the synthetic division array.

1. 5 | $\begin{array}{rrrr} 1 & -2 & -14 & -5 \\ & 5 & 15 & 5 \\ \hline 1 & 3 & 1 & 0 \end{array}$

2. −2 | $\begin{array}{rrrr} 2 & 3 & 3 & 17 \\ & -4 & 2 & -10 \\ \hline 2 & -1 & 5 & 7 \end{array}$

3. 3 | $\begin{array}{rrrr} 1 & 0 & 1 & -2 \\ & 3 & 9 & 30 \\ \hline 1 & 3 & 10 & 28 \end{array}$

Divide using polynomial long division.

4. $(x^2 + 3x - 6) \div (x + 1)$

5. $(x^2 + x - 3) \div (x - 2)$

6. $(x^2 + 5x + 2) \div (x + 3)$

7. $(x^2 - 7x + 1) \div (x - 1)$

8. $(x^2 + 5x + 6) \div (x + 3)$

9. $(x^2 - 3x - 5) \div (x - 5)$

10. $(x^2 - 2x - 8) \div (x - 4)$

11. $(x^2 - 3x + 1) \div (x + 1)$

12. $(x^2 - 5x + 3) \div (x - 2)$

Divide using synthetic division.

13. $(x^2 + 7x + 4) \div (x + 3)$

14. $(x^2 + 2x + 1) \div (x + 1)$

15. $(x^2 + 3x - 2) \div (x - 2)$

16. $(x^2 + 7x - 9) \div (x + 1)$

17. $(x^2 - 3x + 8) \div (x - 2)$

18. $(x^2 - 7x + 10) \div (x - 5)$

19. $(x^2 + 6x + 3) \div (x + 1)$

20. $(x^2 - 5x - 6) \div (x + 1)$

21. $(x^2 - 3x - 1) \div (x - 4)$

You are given an expression for the area of the rectangle. Find an expression for the missing dimension.

22. $A = x^2 + 10x + 21$

$x + 3$ ⌐ [ ] ⌐ ? ⌐

23. $A = x^2 + 2x - 8$

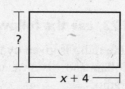

24. $A = x^2 + 8x + 15$

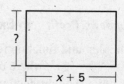

Find the error in the example and correct it.

25. $(x^2 - 6x - 1) \div (x - 3)$

3 | $\begin{array}{rrr} 1 & -6 & -1 \\ & 3 & -9 \\ \hline 1 & -3 & -10 \end{array}$

$x - 3 - \dfrac{10}{x + 3}$

26. $(x^2 - 4x + 5) \div (2x - 3)$

3 | $\begin{array}{rrr} 1 & -4 & 5 \\ & 3 & -3 \\ \hline 1 & -1 & 2 \end{array}$

$x - 1 + \dfrac{2}{2x + 3}$

NAME _____ DATE _____

Practice B

For use with pages 352–358

Divide using polynomial long division.

1. $(x^2 + 2x + 6) \div (x - 3)$

2. $(2x^2 + x - 3) \div (x - 1)$

3. $(x^3 - x^2 - x - 2) \div (x - 2)$

4. $(4x^3 - 7x + 8) \div (2x - 1)$

5. $(3x^3 + 2x^2 - 5x + 1) \div (3x + 1)$

6. $(8x^2 - 5x + 1) \div (2x - 3)$

7. $(x^3 + 5x^2 + 5x - 3) \div (x^2 + 3x - 1)$

8. $(x^3 - 3x^2 + 4x - 6) \div (x^2 + x - 4)$

Divide using synthetic division.

9. $(2x^3 - 7x^2 - x - 12) \div (x - 4)$

10. $(x^3 - 2x + 12) \div (x + 3)$

11. $(x^4 - 5x^3 + 4x - 17) \div (x - 5)$

12. $(3x^3 - 2x^2 + 5x - 1) \div (x + 2)$

13. $(5x^4 - 2x^3 - 3x^2 + 5x + 1) \div (x - 1)$

14. $(3x^4 + 2x^3 - 5) \div (x + 4)$

15. $(x^3 - 2) \div (x + 1)$

16. $(3x^4 - 1) \div (x - 2)$

Given one zero of the polynomial function, find the other zeros.

17. $f(x) = x^3 + 3x^2 - 34x + 48;\ 3$

18. $f(x) = x^3 + 2x^2 - 20x + 24;\ -6$

19. $f(x) = 2x^3 + 3x^2 - 3x - 2;\ -2$

20. $f(x) = 3x^3 - 16x^2 + 3x + 10;\ 5$

21. *Geometry* The volume of the box shown below is given
by $V = 2x^3 - 11x^2 + 10x + 8$. Find an expression for the
missing dimension.

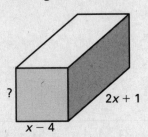

?   2x + 1   x − 4

Company Profit **In Exercises 22 and 23, use the following information.**

The demand function for a type of portable radio is given by the model
$p = 70 - 5x^2$, where p is measured in dollars and x is measured in millions
of units. The production cost is \$20 per radio.

22. Write an equation giving profit as a function of x million radios sold.

23. The company currently produces 3 million radios and makes a profit
of \$15,000,000, but would like to scale back production. What lesser
number of radios could the company produce to yield the same profit?

NAME _____ DATE _____

Practice C

For use with pages 352–358

Divide using polynomial long division.

1. $(x^3 - 3x^2 + 2x - 6) \div (x^2 + 3x - 1)$

2. $(4x^3 - 2x^2 + 6x - 1) \div (2x + 3)$

3. $(2x^4 + 3x - 1) \div (x^2 + 2x + 1)$

4. $(4x^3 - 6x^2 + 5) \div (x^2 - 4)$

5. $(6x^3 - 2x^2 + 5) \div (3x^2 + x)$

6. $(x^3 - 2x^2 + 5x - 1) \div (3x + 2)$

7. $(x^3 + 5) \div (2x^2 - 1)$

8. $(2x^3 - 4x^2 + 3x + 5) \div (4x^2 + 2x - 1)$

Divide using synthetic division.

9. $(5x^4 - 2x^3 + 7x^2 + 6x - 8) \div (x - 2)$

10. $(6x^3 - 2x^2 + 5x + 3) \div (x + 3)$

11. $(2x^3 - 3x + 4) \div (x - 1)$

12. $(4x^3 - 2x^2 + 1) \div (x + 2)$

13. $(3x^5 + 2x^3 - 5x + 1) \div (x - 3)$

14. $(4x + 2x^3 + 7x^2 - 1) \div (x + 1)$

Given one zero of the polynomial function, find the other zeros.

15. $f(x) = x^3 - 8x^2 + 5x + 14;\ 2$

16. $f(x) = 12x^3 + 8x^2 - 13x + 3;\ \frac{1}{2}$

17. $f(x) = x^3 + x^2 - 13x + 3;\ 3$

18. $f(x) = 2x^3 + 11x^2 + 9x + 2;\ -\frac{1}{2}$

Given two zeros of the polynomial function, find the other zeros.

19. $f(x) = x^4 + 6x^3 - 4x^2 - 54x - 45;\ -3, 3$

20. $f(x) = x^4 - 3x^3 + 3x - 1;\ -1, 1$

21. $f(x) = 2x^4 - 9x^3 + 4x^2 + 21x - 18;\ 2, 3$

22. $f(x) = 3x^4 - 2x^3 - 12x^2 + 6x + 9;\ -\sqrt{3}, \sqrt{3}$

23. $f(x) = x^4 + 2x^3 - 14x^2 - 32x - 32;\ -4, 4$

24. $f(x) = x^4 + 3x^3 + 7x^2 + 15x + 10;\ -2, -1$

25. *Hydroelectric Power*   The amount of conventional hydroelectric power (in quadrillion Btu) consumed from 1990 to 1997 can be modeled by $P = -0.004t^3 + 0.082t^2 - 0.268t + 3.206$ where t is the number of years since 1990. For the same years, the U.S. population (in millions) can be modeled by $P = 2.61t + 247$ where t is the number of years since 1990. Find a function for the average amount of energy consumed by each person from 1990 to 1997. What was the per capita consumption of conventional power in 1992?

26. *Yearbook Sales*   If the school charges $15 for a yearbook, 800 students will buy a yearbook. For every $.50 reduction in price two more books are sold. It costs $10 to produce each book. How many books must be sold to earn a profit of at least $2000?

Reteaching with Practice

For use with pages 352–358

GOAL How to divide polynomials and relate the result to the remainder theorem and the factor theorem

VOCABULARY

In **polynomial long division,** when you divide a polynomial $f(x)$ by a divisor $d(x)$, you get a quotient polynomial $q(x)$ and a remainder polynomial $r(x)$. This can be written as $\dfrac{f(x)}{d(x)} = q(x) + \dfrac{r(x)}{d(x)}$. According to the remainder theorem, if a polynomial $f(x)$ is divided by $x - k$, then the remainder is $r = f(k)$.

In **synthetic division,** you only use the coefficients of the polynomial and the divisor must be in the form $x - k$. According to the factor theorem, a polynomial $f(x)$ has a factor $x - k$ if and only if $f(k) = 0$.

EXAMPLE 1 *Using Polynomial Long Division*

Divide $x^4 - 3x^3 - 2x + 1$ by $x^2 + 1$.

SOLUTION

$$
\begin{array}{r}
x^2 - 3x - 1 \\
x^2 + 1 \overline{)x^4 - 3x^3 + 0x^2 - 2x + 1}
\end{array}
$$

Include $0x^2$ for the missing term.

$$\underline{x^4 \qquad + x^2} \qquad\qquad x^2(x^2 + 1)$$

$$-3x^3 - x^2 - 2x \qquad\qquad \text{Subtract.}$$

$$\underline{-3x^3 \qquad - 3x} \qquad\qquad -3x(x^2 + 1)$$

$$-x^2 + x + 1 \qquad\qquad \text{Subtract.}$$

$$\underline{-x^2 \qquad - 1} \qquad\qquad -1(x^2 + 1)$$

$$x + 2 \qquad\qquad \text{remainder}$$

$$\dfrac{x^4 - 3x^3 - 2x + 1}{x^2 + 1} = x^2 - 3x - 1 + \dfrac{x + 2}{x^2 + 1}$$

Exercises for Example 1

Divide using polynomial long division.

1. $(2x^2 - 3x + 1) \div (x + 5)$ **2.** $(3x^2 - x + 4) \div (x - 1)$

3. $(x^2 - 2x + 6) \div (x + 2)$ **4.** $(x^2 - 5x - 7) \div (x - 8)$

Reteaching with Practice

For use with pages 352–358

EXAMPLE 2 ## Using Synthetic Division

Divide $x^3 - 10x - 24$ by $x - 3$.

SOLUTION

Since $x - 3$ is in the form $x - k$, $k = 3$.

$$
\begin{array}{r|rrrr}
3 & 1 & 0 & -10 & -24 \\
  &   & 3 & 9 & -3 \\
\hline
  & 1 & 3 & -1 & \boxed{-27}
\end{array}
$$
Include a 0 for the x^2-term.

$\longleftarrow$ — remainder

$$\frac{x^3 - 10x - 24}{x - 3} = x^2 + 3x - 1 + \frac{-27}{x - 3}$$

Exercises for Example 2

Divide using synthetic division.

5. $(x^2 - 4x + 3) \div (x - 1)$

6. $(2x^2 - 5x - 3) \div (x - 3)$

7. $(x^2 + 2x - 15) \div (x + 4)$

8. $(3x^2 + 10x + 8) \div (x + 2)$

EXAMPLE 3 ## Factoring a Polynomial

Factor $f(x) = 3x^3 - 11x^2 - 6x + 8$ given that $f(4) = 0$.

SOLUTION

Because $f(4) = 0$, $x - 4$ is a factor of $f(x)$. Use synthetic division to find the other factors.

$$
\begin{array}{r|rrrr}
4 & 3 & -11 & -6 & 8 \\
  &   & 12 & 4 & -8 \\
\hline
  & 3 & 1 & -2 & 0
\end{array}
$$

$3x^3 - 11x^2 - 6x + 8 = (x - 4)(3x^2 + x - 2)$ Write $f(x)$ as the product of two factors.

$= (x - 4)(3x - 2)(x + 1)$ Factor trinomial.

Exercises for Example 3

Factor the polynomial given that $f(k) = 0$.

9. $f(x) = x^3 + 2x^2 - 9x - 18$; $k = 3$

10. $f(x) = x^3 + x^2 - 10x + 8$; $k = -4$

NAME _____ DATE _____

Quick Catch-Up for Absent Students

For use with pages 352–358

The items checked below were covered in class on (date missed) _____

Lesson 6.5: The Remainder and Factor Theorems

_____ **Goal 1:** Divide polynomials and relate the result to the remainder theorem and the factor theorem. (pp. 352–354)

Material Covered:

_____ Example 1: Using Polynomial Long Division

_____ Activity: Investigating Polynomial Division

_____ Student Help: Study Tip

_____ Example 2: Using Synthetic Division

_____ Example 3: Factoring a Polynomial

_____ Example 4: Finding Zeros of a Polynomial Function

Vocabulary:

polynomial long division, p. 352   synthetic division, p. 353

_____ **Goal 2:** Use polynomial division in real-life problems. (p. 355)

Material Covered:

_____ Example 5: Using Polynomial Models

_____ Other (specify) _____

_____

Homework and Additional Learning Support

_____ Textbook (specify) _pp. 356–358_____

_____

_____ Internet: Extra Examples at www.mcdougallittell.com

_____ *Reteaching with Practice* worksheet (specify exercises)_____

_____ *Personal Student Tutor* for Lesson 6.5

Real-Life Application: When Will I Ever Use This?

For use with pages 352–358

Economic Models

Economic models are used by both large and small businesses. Suppose you work in the marketing department of a worldwide manufacturer and your company has recently released a new product on the market.

The demand function p shown at the right for the new product is

$$p = 34 - 2x^2$$

where x is the number of units produced in millions. The cost of producing the new product is $9 per unit.

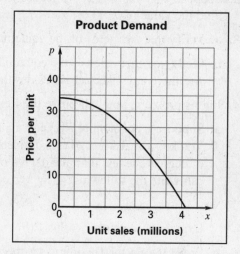

Product Demand

Price per unit

Unit sales (millions)

1. What conclusion can you make about the graph of the demand function?

2. Write an algebraic model that gives your company's profit P as a function of the number of units produced.

3. Sketch the graphs of the revenue, cost, and profit functions.

4. By selling 2.5 million units, the company made a profit of $31.25 million. What lesser number of units could the company have produced to yield the same profit?

5. In Exercise 4, how much was each unit originally being sold for?

6. Sketch the graph of the profit function and show that there are two production levels that produce a profit of $31.25 million.

7. Use a graphing utility to determine the company's profit if 2.8 million units are produced.

8. Your boss wants to increase profits and asks you to determine a unit price that will yield a profit of $34 million. Is this possible? Explain.

NAME _____ DATE _____

Challenge: Skills and Applications

For use with pages 352–358

1. Suppose $x + 1$ is a factor of $f(x) = x^7 + ax - 2$. Find $f(2)$.

2. Suppose $x^3 + 3x^2 - bx + 5$ leaves a remainder of 1 when divided by $x - 2$. Find the value of b.

3. **a.** Verify that two zeros of the equation $x^4 + x^3 - 5x^2 + x - 6 = 0$ are i and $-i$.

   b. Find two other zeros of the equation in part (a), using the Factor Theorem. Are there any other zeros besides these four?

4. In this problem, you will prove the Remainder Theorem and the Factor Theorem.

   a. Suppose a polynomial $P(x)$ and a number b (which could be complex) are given. You know that when $P(x)$ is divided by $x - b$, the quotient is a polynomial $Q(x)$ of degree one less than the degree of $P(x)$ and the remainder is a number R; so you can write the following:

   $$P(x) = (x - b)Q(x) + R$$

   Explain how you know that $P(b) = R$. What does this prove?

   b. Prove the Factor Theorem as a special case of the result of part (a).

5. **a.** If we want to perform a long division such as $1 \div (1 - x)$, where the denominator has higher degree than the numerator, we can write the divisor *from lowest power to highest power*:

$$
\begin{array}{r}
1 \\
1 - x \overline{)\,1 } \\
\underline{1 - x} \\
x
\end{array}
$$

   Continue this long division to get an "infinitely long polynomial" (called a *power series*) as the quotient. Write your result as an equation.

   b. Evaluate the two sides of the equation you wrote in part (a) for $x = \frac{1}{2}$. (You can only evaluate a few terms of the quotient, of course.) Are they approximately equal?

   c. Find a power series representation for the quotient $\dfrac{1}{(1 - x)^2}$.

Algebra 2
Chapter 6 Resource Book

TEACHER'S NAME _____ CLASS _____ ROOM _____ DATE _____

Lesson Plan

1-day lesson (See *Pacing the Chapter,* TE pages 320C–320D)   **For use with pages 359–365**

 GOALS   1. **Find the rational zeros of a polynomial function.**
2. **Use polynomial equations to solve real-life problems.**

State/Local Objectives _____

_____

✓ Check the items you wish to use for this lesson.

STARTING OPTIONS

_____ Homework Check: TE page 356; Answer Transparencies

_____ Warm-Up or Daily Homework Quiz: TE pages 359 and 358, CRB page 77, or Transparencies

TEACHING OPTIONS

_____ Motivating the Lesson: TE page 360

_____ Lesson Opener (Visual Approach): CRB page 78 or Transparencies

_____ Examples 1–3: SE pages 359–361

_____ Extra Examples: TE pages 360–361 or Transparencies

_____ Closure Question: TE page 361

_____ Guided Practice Exercises: SE page 362

APPLY/HOMEWORK

Homework Assignment

_____ Basic   16–30 even, 34–40 even, 46–48, 59, 65–66, 71–83 odd; Quiz 2: 1–19

_____ Average   16–30 even, 34-50 even, 59–61, 65–66, 71–83 odd; Quiz 2: 1–19

_____ Advanced   16–54 even, 60–64 even, 65–70, 71–83 odd; Quiz 2: 1–19

Reteaching the Lesson

_____ Practice Masters: CRB pages 79–81 (Level A, Level B, Level C)

_____ Reteaching with Practice: CRB pages 82–83 or Practice Workbook with Examples

_____ Personal Student Tutor

Extending the Lesson

_____ Applications (Interdisciplinary): CRB page 85

_____ Challenge: SE page 364; CRB page 86 or Internet

ASSESSMENT OPTIONS

_____ Checkpoint Exercises: TE pages 360–361 or Transparencies

_____ Daily Homework Quiz (6.6): TE page 364, CRB page 90, or Transparencies

_____ Standardized Test Practice: SE page 364; TE page 364; STP Workbook; Transparencies

_____ Quiz (6.4–6.6): SE page 365; CRB page 87

Notes _____

_____

_____

TEACHER'S NAME _____ CLASS _____ ROOM _____ DATE _____

Lesson Plan for Block Scheduling

Half-day lesson (See *Pacing the Chapter,* TE pages 320C–320D)   For use with pages 359–365

GOALS 1. **Find the rational zeros of a polynomial function.**
          2. **Use polynomial equations to solve real-life problems.**

State/Local Objectives _____

_____

_____

CHAPTER PACING GUIDE	
Day	**Lesson**
1	6.1 (all); 6.2(all)
2	6.3 (all)
3	6.4 (all)
4	6.5 (all); **6.6(all)**
5	6.7 (all); 6.8 (all)
6	6.9(all); Review Ch. 6
7	Assess Ch. 6; 7.1 (all)

✓ **Check the items you wish to use for this lesson.**

STARTING OPTIONS

_____ Homework Check: TE page 356; Answer Transparencies

_____ Warm-Up or Daily Homework Quiz: TE pages 359 and 358,
         CRB page 77, or Transparencies

TEACHING OPTIONS

_____ Motivating the Lesson: TE page 360

_____ Lesson Opener (Visual Approach): CRB page 78 or Transparencies

_____ Examples 1–3: SE pages 359–361

_____ Extra Examples: TE pages 360–361 or Transparencies

_____ Closure Question: TE page 361

_____ Guided Practice Exercises: SE page 362

APPLY/HOMEWORK

Homework Assignment (See also the assignment for Lesson 6.5.)

_____ Block Schedule: 16–30 even, 34-50 even, 59–61, 65–66, 71–83 odd; Quiz 2: 1–19

Reteaching the Lesson

_____ Practice Masters: CRB pages 79–81 (Level A, Level B, Level C)

_____ Reteaching with Practice: CRB pages 82–83 or Practice Workbook with Examples

_____ Personal Student Tutor

Extending the Lesson

_____ Applications (Interdisciplinary): CRB page 85

_____ Challenge: SE page 364; CRB page 86 or Internet

ASSESSMENT OPTIONS

_____ Checkpoint Exercises: TE pages 360–361 or Transparencies

_____ Daily Homework Quiz (6.6): TE page 364, CRB page 90, or Transparencies

_____ Standardized Test Practice: SE page 364; TE page 364; STP Workbook; Transparencies

_____ Quiz (6.4–6.6): SE page 365; CRB page 87

Notes _____

_____

_____

NAME _____ DATE _____

WARM-UP EXERCISES

For use before Lesson 6.6, pages 359–365

Factor each polynomial.

1. $6x^2 + 13x - 28$

2. $5x^3 - 22x^2 + 12x - 16$

Find the zeros of each function.

3. $f(x) = 2x^2 + 3x - 2$

4. $f(x) = x^4 - 9x^2 + 20$

5. Find $f\left(\frac{5}{4}\right)$ when $f(x) = 8x^3 - 22x^2 + 47x - 40$.

DAILY HOMEWORK QUIZ

For use after Lesson 6.5, pages 352–358

1. Divide using polynomial long division.
$(x^2 - 7x - 12) \div (x + 5)$

2. Divide using synthetic division.
$(2x^3 + 6x^2 - 9) \div (x + 2)$

3. Factor the polynomial $f(x) = x^3 - 2x^2 - 19x + 20$ given that
$f(5) = 0$.

4. Given that -3 is a zero of $f(x) = x^3 - 6x^2 - 13x + 42$, find
the other zeros.

Available as
a transparency

NAME _____ DATE _____

Visual Approach Lesson Opener

For use with pages 359–365

1. Let $f(x) = \mathbf{63}x^3 - 129x^2 + 28x + 20$. Show that $f(x)$ can be factored as $(\mathbf{3}x - \mathbf{5})(\mathbf{7}x + \mathbf{2})(\mathbf{3}x - \mathbf{2})$.

2. Solve $f(x) = 0$. Use a colored highlighter pen to highlight the numbers 3 and 7 where they occur in your list of solutions.

3. Complete this sentence: The demoninators of the solutions to $f(x) = 0$ are all _____ of the leading coefficient 63 in the original expression for $f(x)$.

4. Now let's concentrate on the constant terms. We may write $f(x) = 63x^3 - 129x^2 + 28x + \mathbf{20}$ or, in factored form, $f(x) = (3x - \mathbf{5})(7x + \mathbf{2})(3x - \mathbf{2})$. Again, write the solution of $f(x) = 0$. This time, highlight the numbers 2 and 5 where they occur.

5. Complete this sentence: The numerators of the solutions to $f(x) = 0$ are all _____ of the constant term 20 in the original expression for $f(x)$.

6. Let $g(x) = 6x^2 + x - 35 = 0$. Solve $g(x) = 0$.

7. Complete these sentences:

The _____ of the solutions to $g(x) = 0$ are all _____ of the leading coefficient _____ in the original expression for $g(x)$.

The _____ of the solutions to $g(x) = 0$ are all _____ of the constant term _____ in the original expression for $g(x)$.

NAME _____ DATE _____

Practice A

For use with pages 359–365

List the possible rational zeros of *f* using the rational zero theorem.

1. $f(x) = x^3 + 2x^2 + 4x + 1$

2. $f(x) = x^2 + 3x + 7$

3. $f(x) = x^3 + 2x^2 - 5x + 6$

4. $f(x) = x^4 - 6x + 9$

5. $f(x) = x^3 + 3x^2 - 12$

6. $f(x) = x^8 - 2x^5 + x^4 - 3x + 20$

7. $f(x) = x^5 + 2x^4 - 3x - 24$

8. $f(x) = x^2 + 6x - 40$

9. $f(x) = x^3 - 5x^2 + 2x + 100$

Use synthetic division to decide which of the following are zeros of the function: 1, −1.

10. $f(x) = x^2 + 6x + 5$

11. $f(x) = x^3 - x^2 - 9x + 9$

12. $f(x) = x^2 + 10x + 21$

13. $f(x) = x^3 + 2x^2 - 2x - 1$

14. $f(x) = x^3 + 3x^2 - 4x - 12$

15. $f(x) = x^3 - x^2 - x + 1$

16. $f(x) = x^3 + 5x^2 - x - 5$

17. $f(x) = x^3 + 6x^2 + 8x$

18. $f(x) = 2x^2 - x - 1$

Find all the rational zeros of the function.

19. $f(x) = x^3 + x^2 - 14x - 24$

20. $f(x) = x^3 - 2x^2 - x + 2$

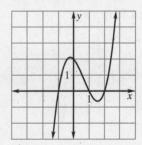

Find all the real zeros of the function.

21. $f(x) = x^3 + 3x^2 - 4x - 12$

22. $f(x) = x^3 + 3x^2 - 3x - 5$

23. $f(x) = x^3 - 3x^2 - 5x + 15$

Geometry   **In Exercises 24–26, use the following information.**

The volume of the box shown at the right is given by $V = x^3 + 5x^2 + 4x$.

24. Write an equation that indicates that the volume of the box is 84 in.^{3}.

25. Use the rational zero theorem to list all possible rational zeros of the equation in Exercise 24.

26. Find the dimensions of the box.

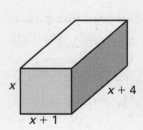

NAME _____ DATE _____

Practice B

For use with pages 359–365

List the possible rational zeros of *f* using the rational zero theorem.

1. $f(x) = x^4 - 2x^3 + 3x - 4$ **2.** $f(x) = 2x^3 - x^2 + 5x + 6$ **3.** $f(x) = 3x^5 + 2x + 8$

Use the rational zero theorem and synthetic division to find all rational zeros of the function.

4. $f(x) = 2x^3 - 3x^2 - 11x + 6$

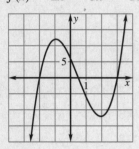

5. $f(x) = 3x^3 + 8x^2 - 3x - 8$

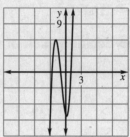

6. $f(x) = 8x^3 - 6x^2 - 23x + 6$ **7.** $f(x) = x^3 - 4x^2 - 7x + 10$

8. $f(x) = x^4 + 4x^3 + x^2 - 8x - 6$ **9.** $f(x) = 2x^4 + 5x^3 - 5x^2 - 5x + 3$

Find all real zeros of the function.

10. $f(x) = 2x^3 - 5x^2 - 4x + 10$

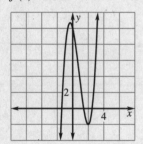

11. $f(x) = 4x^3 - 8x^2 - 15x + 9$

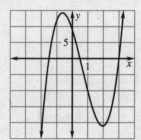

12. $f(x) = x^3 - 3x^2 - 3x + 9$ **13.** $f(x) = 2x^4 + 3x^3 - 6x^2 - 6x + 4$

14. $f(x) = x^4 + 2x^3 - 5x^2 - 4x + 6$ **15.** $f(x) = 2x^4 + 5x^3 - 6x^2 - 7x + 6$

European College Students   **In Exercises 16–19, use the following information.**

Many students from Europe come to the United States for their college education. From 1980 through 1990, the number S (in thousands), of European students attending a college or university in the U.S. can be modeled by $S = 0.07(t^3 - 13t^2 + 65t + 339)$ where $t = 0$ corresponds to 1980.

16. Write an equation with a leading coefficient of 1 that represents the year that 31.08 thousand European students attended a U.S. college or university.

17. Use the rational zero theorem to list all possible rational zeros of the equation in Exercise 16.

18. Which of the rational zeros listed in Exercise 17 are valid values of *t*?

19. In what year did 31.08 thousand European students attend a U.S. college or university?

NAME _____ DATE _____

Practice C

For use with pages 359–365

Use the rational zero theorem and synthetic division to find all rational zeros of the function.

1. $f(x) = 4x^3 - 8x^2 - 29x - 12$

2. $f(x) = 3x^4 + 10x^3 - 11x^2 - 10x + 8$

3. $f(x) = 10x^4 - 43x^3 + 11x^2 + 79x + 15$

4. $f(x) = 24x^4 - 26x^3 - 45x^2 - x + 6$

5. $f(x) = 6x^4 + 35x^3 + 35x^2 - 55x - 21$

6. $f(x) = 8x^3 + 28x^2 + 14x - 15$

Find all real zeros of the function.

7. $f(x) = x^3 + 2x^2 - 34x + 7$

8. $f(x) = 6x^3 - 49x^2 + 20x - 2$

9. $f(x) = 3x^3 + 10x^2 - 7x - 20$

10. $f(x) = x^4 + 4x^3 - 14x^2 - 20x - 3$

11. $f(x) = 6x^4 + 25x^3 + 32x^2 + 15x + 2$

12. $f(x) = 12x^4 + 28x^3 - 11x^2 - 13x + 5$

13. $f(x) = 6x^4 + 31x^3 - 64x^2 - 489x - 540$

14. $f(x) = 8x^4 + 68x^3 + 178x^2 + 103x - 105$

Critical Thinking   **In Exercises 15–18, consider the function**
$f(x) = x^4 - x^3 - 24x^2 - 36x.$

15. Explain why the rational zero theorem cannot be directly applied to this function.

16. Factor out the common monomial factor of f.

17. Apply the rational zero theorem to find all other rational zeros of f.

18. Find all the real zeros of $f(x) = 3x^5 + x^4 - 12x^3 - 4x^2.$

Critical Thinking   **In Exercises 19–22, consider the functions**
$f(x) = x^3 + 2x^2 - x - 2, g(x) = -x^3 - 2x^2 + x + 2,$
$h(x) = 2x^3 + 4x^2 - 2x - 4,$ **and** $j(x) = 5x^3 + 10x^2 - 5x - 10.$

19. Use the rational zero theorem to find all rational zeros of each function.

20. Note that $g(x) = -f(x)$, $h(x) = 2f(x)$, and $j(x) = 5f(x)$. What can you conclude about the zeros of $f(x)$ and $af(x)$?

21. Explain why the rational zero theorem cannot be directly applied to $f(x) = \frac{1}{2}x^3 - \frac{19}{2}x - 15.$

22. Use the conclusion from Exercise 20 to find the rational zeros of the function in Exercise 21.

NAME _____ DATE _____

Reteaching with Practice

For use with pages 359–365

GOAL **How to find the rational zeros of a polynomial function**

> **VOCABULARY**
>
> According to the rational zero theorem, if $f(x) = a_n x^n + \cdots + a_1 x + a_0$ has *integer* coefficients, then every rational zero of f has the form:
>
> $\dfrac{p}{q} = \dfrac{\text{factor of constant term } a_0}{\text{factor of leading coefficient } a_n}$

EXAMPLE 1 *Using the Rational Zero Theorem*

Find the rational zeros of $f(x) = 3x^3 - 4x^2 - 17x + 6$.

SOLUTION

Begin by listing the possible rational zeros. The leading coefficient is 3 and the constant term is 6

$$x = \frac{\pm 1, \pm 2, \pm 3, \pm 6}{\pm 1, \pm 3} \qquad \begin{array}{l}\text{Factors of constant term, 6} \\ \text{Factors of leading coefficient, 3}\end{array}$$

Test these zeros using synthetic division.

Test $x = 1$:

1	3	−4	−17	6
		3	−1	−18
	3	−1	−18	−12

Test $x = -1$:

−1	3	−4	−17	6
		−3	7	10
	3	−7	−10	16

Test $x = 2$:

2	3	−4	−17	6
		6	4	−26
	3	2	−13	−20

Test $x = -2$:

−2	3	−4	−17	6
		−6	20	−6
	3	−10	3	0

Since the remainder for $x = -2$ is 0, $x + 2$ is a factor of f.

$f(x) = (x + 2)(3x^2 - 10x + 3)$

$\qquad = (x + 2)(3x - 1)(x - 3)$ Factor trinomial.

The zeros of f are $-2, \frac{1}{3}$, and 3.

Exercises for Example 1
. . . .

Find the rational zeros of the function.

1. $f(x) = x^3 - 7x + 6$

2. $f(x) = 2x^3 + 2x^2 - 8x - 8$

3. $f(x) = x^3 + x^2 - 10x + 8$

4. $f(x) = 2x^3 - 3x^2 - 8x - 3$

NAME _____ DATE _____

Reteaching with Practice

For use with pages 359–365

Lesson 6.6

EXAMPLE 2 *Using the Rational Zero Theorem*

Find all real zeros of $f(x) = 3x^4 + x^3 - 8x^2 - 2x + 4$.

SOLUTION

Begin by listing the possible rational zeros.

$$x = \frac{\pm 1, \pm 2, \pm 4}{\pm 1, \pm 3} \quad \begin{array}{l} \text{Factors of constant term, 4.} \\ \text{Factors of leading coefficient, 3.} \end{array}$$

Then test these zeros using synthetic division.

Test $x = 1$:

$$\begin{array}{r|rrrrr} 1 & 3 & 1 & -8 & -2 & 4 \\ & & 3 & 4 & -4 & -6 \\ \hline & 3 & 4 & -4 & -6 & -2 \end{array}$$

Test $x = -1$:

$$\begin{array}{r|rrrrr} -1 & 3 & 1 & -8 & -2 & 4 \\ & & -3 & 2 & 6 & -4 \\ \hline & 3 & -2 & -6 & 4 & 0 \end{array}$$

Since the remainder for $x = -1$ is 0, $x + 1$ is a factor of f.

$$f(x) = (x + 1)(3x^3 - 2x^2 - 6x + 4)$$

Repeat this process for $g(x) = 3x^3 - 2x^2 - 6x + 4$. The possible rational zeros are $\pm 1, \pm 2, \pm 4, \pm \frac{1}{3}, \pm \frac{2}{3}, \pm \frac{4}{3}$.

To save time, graph the function f and estimate where it crosses the x-axis. A reasonable choice from the list of possibilities is $x = \frac{2}{3}$.

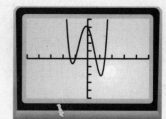

$$\begin{array}{r|rrrr} \frac{2}{3} & 3 & -2 & -6 & 4 \\ & & 2 & 0 & -4 \\ \hline & 3 & 0 & -6 & 0 \end{array}$$

Again, since the remainder is 0, $x - \frac{2}{3}$ is a factor of g.

So, $f(x) = (x + 1)\left(x - \frac{2}{3}\right)(3x^2 - 6)$

$\quad\quad = 3(x + 1)\left(x - \frac{2}{3}\right)(x^2 - 2)$ $\quad$ Factor out a 3.

$\quad\quad = 3(x + 1)\left(x - \frac{2}{3}\right)\left(x + \sqrt{2}\right)\left(x - \sqrt{2}\right)$ $\quad$ Difference of two squares

The real zeros of f are $-1, \frac{2}{3}, -\sqrt{2}$, and $\sqrt{2}$.

Exercises for Example 2
....

Find all the real zeros of the function.

5. $f(x) = x^3 + 3x^2 - 5x - 15$

6. $f(x) = x^4 - \frac{3}{2}x^3 - 7x^2 + 9x + 6$

7. $f(x) = x^3 - 5x^2 + 5x - 1$

8. $f(x) = x^3 - x^2 - 3x - 1$

NAME _____ DATE _____

Quick Catch-Up for Absent Students

For use with pages 359–365

The items checked below were covered in class on (date missed) _____

Lesson 6.6: Finding Rational Zeros

_____ **Goal 1:** Find the rational zeros of a polynomial function. (pp. 359, 360)

Material Covered:

_____ Example 1: Using the Rational Zero Theorem

_____ Example 2: Using the Rational Zero Theorem

_____ **Goal 2:** Use polynomial equations to solve real-life problems. (p. 361)

Material Covered:

_____ Example 3: Writing and Using a Polynomial Model

_____ Other (specify) _____

_____

Homework and Additional Learning Support

_____ Textbook (specify) _pp. 362–365_____

_____

_____ Internet: Extra Examples at www.mcdougallittell.com

_____ *Reteaching with Practice* worksheet (specify exercises)_____

_____ *Personal Student Tutor* for Lesson 6.6

NAME _____ DATE _____

Interdisciplinary Application

For use with pages 359–365

Aquariums

BIOLOGY   Aquariums can range in size anywhere from a 20-gallon home aquarium to large buildings that contain exhibits of aquatic wildlife. Large tanks, called oceanariums, can hold more than 5 million gallons of water! Public aquariums are popular tourist attractions that are visited by millions of people each year. They also can serve as important research, education, and conservation centers. These public aquariums are fascinating in the fact that only a pane of glass can separate an observer from a killer whale or a shark.

Aquarists, marine biologists, and mammologists are just some of the people that help maintain public aquariums. It is their responsibility to not only maintain the living conditions inside the aquarium but to feed and nourish the wildlife that live there.

In Exercises 1 and 2, use the following information.

The Tennessee Aquarium, in Chattanooga, Tennessee, is the largest freshwater aquarium in the world with 450,000 gallons of water in 24 exhibits. The 130,000 square feet of the aquarium are home to more than 9000 creatures. An interesting fact is that over half a million crickets, about 400,000 worms, and 15,000 pounds of seafood are devoured by the aquarium's wildlife each year.

1. How many gallons of water, on average, does each exhibit in the aquarium have?

2. Suppose the Tennessee Aquarium is opening a new saltwater exhibit that will have tropical fish from around the world. Three of the new rectangular aquariums being ordered are described below.

Aquarium 1 (180-gallon tank): The aquarium has a length of x inches, a height and width that are both 48 inches less than the length, and a volume of 41,472 cubic inches. Write a polynomial in standard form to describe the volume of the aquarium. Then find its dimensions.

Aquarium 2 (300-gallon tank): The aquarium has dimensions of x feet long, $x - 6$ feet wide, and $x - 5.5$ feet high. The volume of the aquarium is 40 cubic feet. Write a polynomial in standard form to describe the volume of the aquarium. Then find its dimensions.

Aquarium 3 (125-gallon tank): The length of this aquarium is three times its height, and its width is 4 inches more than the height. The volume of the aquarium is 28,800 cubic inches. Write a polynomial in standard form to describe the volume of the aquarium. Then find its dimensions.

Challenge: Skills and Applications

For use with pages 359–365

1. Suppose that the coefficients of the equation $P(x) = ax^3 + bx^2 + cx + d = 0$ satisfy the equation $a + b + c + d = 0$. Name a zero of the equation.

2. Repeat Exercise 1, but with the coefficients satisfying the equation $-a + b - c + d = 0$.

3. **a.** Suppose i is a zero of the equation $P(x) = 0$ in Exercise 1. State two relationships that must hold among the coefficients a, b, c, and/or d.

   b. Suppose, as in part (a), that i is a zero of the equation $P(x) = 0$, and the two relationships you found are true. Must $-i$ also be a zero of the equation?

4. **a.** Suppose 3 is a zero of $ax^4 - 2x^3 - 5x^2 + 4x + d = 0$, and suppose $|d| < 10$. List all the possible values of d.

   b. Let $a = 1$ in the equation in part (a). Find the value of d, and find the other zeros of the equation.

5. One question that concerned ancient Greek mathematicians was that of *doubling the cube*, that is, how to construct the length x of a side of a cube whose volume is double that of a given cube, say of side 1. This number x would therefore satisfy the equation $x^3 = 2$, or $x^3 - 2 = 0$.

   a. List the possible rational zeros of this equation.

   b. Explain how you know that any number x that satisfies this equation must be irrational.

6. In this exercise you will prove the rational zero theorem for the case in which $f(x) = ax^3 + bx^2 + cx + d$, and p and q are both prime numbers ($p \neq q$).

   a. Suppose $\frac{p}{q}$ is a zero of $f(x) = 0$. Substitute this value for x, and multiply both sides of this equation by q^3. Solve the resulting equation for ap^3.

   b. Use the equation you found in part (a) to explain how you know that a must be evenly divisible by q. Use the fact that if a product of two numbers mn is divisible by a prime number q, then either m or n (or both) must be divisible by q.

   c. By solving the equation you wrote in part (a) for dq^3, show similarly that d must be evenly divisible by p.

NAME _____ DATE _____

Quiz 2

For use after Lessons 6.4–6.6

1. Factor the polynomial $3x^3 + 81$. *(Lesson 6.4)*

2. Factor the polynomial $4x^3 + 12x^2 + 12x + 36$. *(Lesson 6.4)*

3. Factor the polynomial $5x^4 - 45$. *(Lesson 6.4)*

4. Find the real-number solutions of the equation $x^4 - 9x^2 + 20 = 0$. *(Lesson 6.4)*

5. Divide $x^3 - 3x^2 + 3x - 6$ by $x - 5$ using synthetic division. *(Lesson 6.5)*

6. Divide $8x^2 - 10x - 7$ by $4x + 1$. *(Lesson 6.5)*

7. Divide $x^4 + 6x^3 - 10x^2 + 2x - 1$ by $x^3 + 5$. *(Lesson 6.5)*

8. Find all the real zeros of the function $f(x) = 2x^3 - x^2 - 7x + 6$. *(Lesson 6.6)*

Answers

1. _____

2. _____

3. _____

4. _____

5. _____

6. _____

7. _____

8. _____

TEACHER'S NAME _____ CLASS _____ ROOM _____ DATE _____

Lesson Plan

1-day lesson (See *Pacing the Chapter,* TE pages 320C–320D)   **For use with pages 366–372**

1. **Use the fundamental theorem of algebra to determine the number of zeros of a polynomial function.**
2. **Use technology to approximate the real zeros of a polynomial function.**

State/Local Objectives _____

_____

✓ Check the items you wish to use for this lesson.

STARTING OPTIONS
____ Homework Check: TE page 362; Answer Transparencies
____ Warm-Up or Daily Homework Quiz: TE pages 366 and 364, CRB page 90, or Transparencies

TEACHING OPTIONS
____ Motivating the Lesson: TE page 367
____ Lesson Opener (Graphing Calculator): CRB page 91 or Transparencies
____ Graphing Calculator Activity with Keystrokes: CRB pages 92–93
____ Examples 1–5: SE pages 366–368
____ Extra Examples: TE pages 367–368 or Transparencies; Internet
____ Technology Activity: SE page 372
____ Closure Question: TE page 368
____ Guided Practice Exercises: SE page 369

APPLY/HOMEWORK
Homework Assignment
____ Basic   16–28 even, 35–38, 47–50, 55–56, 60, 65–71 odd
____ Average   16–30 even, 36–60 even, 65–71 odd
____ Advanced   16–60 even, 61–63, 65–71 odd

Reteaching the Lesson
____ Practice Masters: CRB pages 94–96 (Level A, Level B, Level C)
____ Reteaching with Practice: CRB pages 97–98 or Practice Workbook with Examples
____ Personal Student Tutor

Extending the Lesson
____ Applications (Real-Life): CRB page 100
____ Challenge: SE page 371; CRB page 101 or Internet

ASSESSMENT OPTIONS
____ Checkpoint Exercises: TE pages 367–368 or Transparencies
____ Daily Homework Quiz (6.7): TE page 371, CRB page 104, or Transparencies
____ Standardized Test Practice: SE page 371; TE page 371; STP Workbook; Transparencies

Notes _____

_____

_____

Algebra 2
Chapter 6 Resource Book

Lesson Plan for Block Scheduling

Half-day lesson (See *Pacing the Chapter*, TE pages 320C–320D)   For use with pages 366–372

GOALS

1. **Use the fundamental theorem of algebra to determine the number of zeros of a polynomial function.**
2. **Use technology to approximate the real zeros of a polynomial function.**

State/Local Objectives _____

_____

_____

✓ **Check the items you wish to use for this lesson.**

STARTING OPTIONS

____ Homework Check: TE page 362; Answer Transparencies
____ Warm-Up or Daily Homework Quiz: TE pages 366 and 364, CRB page 90, or Transparencies

TEACHING OPTIONS

____ Motivating the Lesson: TE page 367
____ Lesson Opener (Graphing Calculator): CRB page 91 or Transparencies
____ Graphing Calculator Activity with Keystrokes: CRB pages 92–93
____ Examples 1–5: SE pages 366–368
____ Extra Examples: TE pages 367–368 or Transparencies; Internet
____ Technology Activity: SE page 372
____ Closure Question: TE page 368
____ Guided Practice Exercises: SE page 369

APPLY/HOMEWORK

Homework Assignment (See also the assignment for Lesson 6.8.)
____ Block Schedule: 16–30 even, 36–60 even, 65–71 odd

Reteaching the Lesson
____ Practice Masters: CRB pages 94–96 (Level A, Level B, Level C)
____ Reteaching with Practice: CRB pages 97–98 or Practice Workbook with Examples
____ Personal Student Tutor

Extending the Lesson
____ Applications (Real Life): CRB page 100
____ Challenge: SE page 371; CRB page 101 or Internet

ASSESSMENT OPTIONS

____ Checkpoint Exercises: TE pages 367–368 or Transparencies
____ Daily Homework Quiz (6.7): TE page 371, CRB page 104, or Transparencies
____ Standardized Test Practice: SE page 371; TE page 371; STP Workbook; Transparencies

Notes _____

_____

_____

CHAPTER PACING GUIDE	
Day	Lesson
1	6.1 (all); 6.2(all)
2	6.3 (all)
3	6.4 (all)
4	6.5 (all); 6.6(all)
5	**6.7 (all)**; 6.8 (all)
6	6.9(all); Review Ch. 6
7	Assess Ch. 6; 7.1 (all)

Lesson 6.7

NAME _____ DATE _____

WARM-UP EXERCISES

For use before Lesson 6.7, pages 366–372

Name the rational zeros of each function.

1. $f(x) = x^2 - 6x + 9$

2. $f(x) = 2x^2 - x - 15$

3. $f(x) = x^3 + x^2 + 7x - 9$

4. $f(x) = 2x^3 + x^2 + x - 1$

5. $f(x) = x^4 - x^2 + 2x + 4$

···

DAILY HOMEWORK QUIZ

For use after Lesson 6.6, pages 359–365

1. List the possible rational zeros of $f(x) = 6x^3 + 7x^2 + 3x - 4$.

2. Use synthetic division to decide which of the numbers $2, -2,$ $3,$ and -3 are zeros of $f(x) = 2x^3 + 4x^2 - 18x - 36$.

Find all the real zeros of the function.

3. $f(x) = 2x^3 + 3x^2 - 10x - 15$

4. $f(x) = 6x^3 - 37x^2 + 37x - 10$

Algebra 2
Chapter 6 Resource Book

Graphing Calculator Lesson Opener

For use with pages 366–371

Use the equation $x^3 - 2x^2 - 4x + 8 = 0$, or $(x + 2)(x - 2)^2 = 0$. Since the factor $x - 2$ occurs twice in the factored form, we say $x = 2$ is a *repeated solution.* Counting repeated solutions, the equation has *three* real solutions.

You can also use a graph to recognize repeated solutions. At repeated solutions, the graph only *touches* the x-axis. At other solutions, it *crosses* the x-axis.

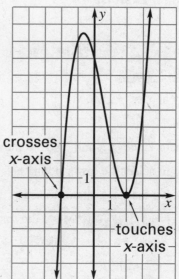

Use a graph and/or factoring to find the number of real solutions (counting repeated solutions) for $f(x) = 0$**.**

1. $f(x) = 2x - 5$

2. $f(x) = 3x + 1$

3. $f(x) = x^2 - 3x + 4$

4. $f(x) = 4x^2 + 20x + 25$

5. $f(x) = x^2 - 2x - 3$

6. $f(x) = x^3 - 2x^2 + 3$

7. $f(x) = x^3 - x^2 - 4x + 4$

8. $f(x) = x^3 - 3x^2 + 4$

9. $f(x) = x^4 - 3x^2 - 4$

10. $f(x) = x^4 - 5x^2 + 4$

Use your results above to guess the possible number of solutions for $f(x) = 0$**, where** $f(x)$ **is the indicated type of function.**

11. Linear function (Exercises 1 and 2)

12. Quadratic function (Exercises 3–5)

13. Cubic function (Exercises 6–8)

14. 4th-degree polynomial (Exercises 9 and 10)

15. Polynomial of degree n (n even)

16. Polynomial of degree n (n odd)

Graphing Calculator Activity Keystrokes

For use with page 370

Keystrokes for Exercise 47

TI-82

`Y=` `X,T,θ` `^` 3 `−` `X,T,θ` `x²` `−` 5
`X,T,θ` `+` 3 `ZOOM` 6

Find zero near $x \approx -2.09$.

`2nd` [CALC] 2

Use the cursor keys, `◄` and `►`, to move the trace cursor to select the lower bound at $x \approx -3$.

`ENTER` Move the trace cursor to select the upper bound at $x \approx -1.5$. `ENTER` Move the trace cursor to select the guess at $x \approx -2$. `ENTER`

Find zero near $x \approx 0.57$.

`2nd` [CALC] 2

Select the lower bound at $x \approx 0$. `ENTER`

Select the upper bound at $x \approx 1$. `ENTER`

Select the guess at $x \approx 0.5$. `ENTER`

Find zero near $x \approx 2.51$.

`2nd` [CALC] 2

Select the lower bound at $x \approx 2$. `ENTER`

Select the upper bound at $x \approx 3$. `ENTER`

Select the guess at $x \approx 2.5$. `ENTER`

TI-83

Step 1:

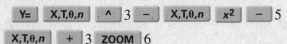

`Y=` `X,T,θ,n` `^` 3 `−` `X,T,θ,n` `x²` `−` 5
`X,T,θ,n` `+` 3 `ZOOM` 6

Find zero near $x \approx -2.09$.

`2nd` [CALC] 2 `(-)` 3 `ENTER` (−1.5) `ENTER` −2 `ENTER`

Find zero near $x \approx 0.57$.

`2nd` [CALC] 2 0 `ENTER` 1 `ENTER` 0.5 `ENTER`

Find zero near $x \approx 2.51$.

`2nd` [CALC] 2 2 `ENTER` 3 `ENTER` 2.5 `ENTER`

SHARP EL-9600c

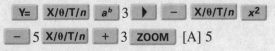

`Y=` `X/θ/T/n` `aᵇ` 3 `►` `−` `X/θ/T/n` `x²`
`−` 5 `X/θ/T/n` `+` 3 `ZOOM` [A] 5

Find zeros of the function.

`2ndF` [CALC] 5

`2ndF` [CALC] 5

`2ndF` [CALC] 5

CASIO CFX-9850GA PLUS

From the main menu, choose GRAPH.

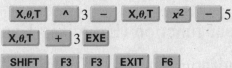

`X,θ,T` `^` 3 `−` `X,θ,T` `x²` `−` 5
`X,θ,T` `+` 3 `EXE`

`SHIFT` `F3` `F3` `EXIT` `F6`

Find zeros of the function.

`SHIFT` `F5` `F1`

Use the cursor keys, `◄` or `►`, to find the other zeros.

Graphing Calculator Activity Keystrokes

For use with page 372

TI-82

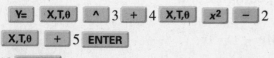

Y= | X,T,θ | ^ 3 + 4 X,T,θ | x² | − 2
X,T,θ | + 5 ENTER

19 ENTER

ZOOM 6

WINDOW ENTER (-) 10 ENTER 10 ENTER 1
ENTER (-) 2 ENTER 22 ENTER 1 ENTER
GRAPH

Find point of intersection near $x \approx -3.35$

2nd [CALC] 5 ENTER ENTER

Use the cursor keys to select guess at $x \approx -3.3$.

ENTER

Find point of intersection near $x \approx -2.40$.

2nd [CALC] 5 ENTER ENTER

Select guess at $x \approx -2.5$ ENTER

Find point of intersection near $x \approx 1.74$.

2nd [CALC] 5 ENTER ENTER

Select guess at $x \approx 1.7$. ENTER

TI-83

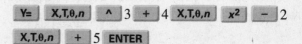

Y= | X,T,θ,n | ^ 3 + 4 X,T,θ,n | x² | − 2
X,T,θ,n | + 5 ENTER

19 ENTER

ZOOM 6

WINDOW (-) 10 ENTER 10 ENTER 1
ENTER (-) 2 ENTER 22 ENTER 1 ENTER 1
ENTER GRAPH

Find point of intersection near $x \approx -3.35$

2nd [CALC] 5 ENTER ENTER (-) 3.3 ENTER

Find point of intersection near $x \approx -2.40$.

2nd [CALC] 5 ENTER ENTER (-) 2.4 ENTER

Find point of intersection near $x \approx 1.74$.

2nd [CALC] 5 ENTER ENTER 1.74 ENTER

SHARP EL-9600c

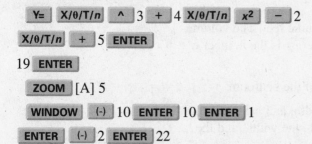

Y= | X/θ/T/n | ^ 3 + 4 X/θ/T/n | x² | − 2
X/θ/T/n | + 5 ENTER

19 ENTER

ZOOM [A] 5

WINDOW (-) 10 ENTER 10 ENTER 1
ENTER (-) 2 ENTER 22
ENTER 1 ENTER GRAPH

Find points of intersection near $x \approx -3.35$,
$x \approx -2.40$, and $x \approx 1.74$.

2ndF [A] 2

2ndF [A] 2

2ndF [A] 2

CASIO CFX-9850GA PLUS

From the main menu, choose GRAPH.

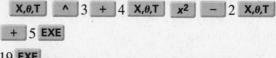

X,θ,T | ^ 3 + 4 X,θ,T | x² | − 2 X,θ,T
+ 5 EXE

19 EXE

SHIFT F3 F3 EXIT F6

SHIFT (-) 10 EXE 10 EXE 1 EXE (-) 2
EXE 22 EXE 1 EXE EXIT F6

Find points of intersection near $x \approx -3.35$,
$x \approx -2.40$, and $x \approx 1.74$.

SHIFT F5 F5

Use the cursor keys, ◀ or ▶, to move the
cursor to the other zeros.

Practice A

For use with pages 366–371

Determine the total number of solutions (including complex and repeated) of the polynomial equation.

1. $x^3 - 3x^2 + 4x + 2 = 0$

2. $x^5 - 3x^2 + 4x + 1 = 0$

3. $2x^4 - 3x^3 + 2x^2 + x - 5 = 0$

4. $3x^6 + 2x^5 + x^4 - 3x^3 + 2x^2 + x - 8 = 0$

5. $5x^3 - 2x^2 + 3x + 1 = 0$

6. $-6x^2 + 3x + 1 = 0$

7. $5x - 7 = 0$

8. $2x^5 - 3x^4 + x - 9 = 0$

Given that $f(x)$ has real coefficients and $x = k$ is a zero, what other number must be a zero of f?

9. $k = 2 + i$

10. $k = 3 - 5i$

11. $k = -6 - 2i$

12. $k = -7 + 3i$

13. $k = \sqrt{2} + i$

14. $k = \sqrt{5} - 4i$

15. $k = 3 + \sqrt{2}i$

16. $k = -2 + \sqrt{3}i$

17. $k = \sqrt{3} - \sqrt{5}i$

Decide whether the given x-value is a zero of the function.

18. $f(x) = x^3 + 2x^2 + 4x - 7, x = 1$

19. $f(x) = x^3 - 3x^2 + 2x + 1, x = 2$

20. $f(x) = x^3 - x^2 + 4x + 3, x = -3$

21. $f(x) = 2x^4 + x^3 - x^2 + 4x + 4, x = -1$

22. $f(x) = x^3 - 2x^2 - 2x - 3, x = 3$

23. $f(x) = x^4 + 2x^3 - 6x^2 + 5x + 2, x = 2$

Write a polynomial function of least degree that has real coefficients, the given zeros, and a leading coefficient of 1.

24. 6

25. 2, -3

26. $-1, -5$

27. $-1, 1, 3$

28. 0, 3, 4

29. $-2, 3, 7$

Room Dimension Riddle **In Exercises 30–32, use the following information.**

One of the bedrooms of a house has a volume of 1144 cubic feet. The volume of the bedroom is given by $y = x^3 + 2x^2 - 5x - 6$, where x is the number of rooms in the house.

30. Factor the polynomial that represents the volume of the bedroom.

31. The factors in Exercise 30 represent the length, width, and height of the bedroom. Which do you think represents the length, the width, and the height?

32. How many rooms does the house have?

NAME _____ DATE _____

Practice B

For use with pages 366–371

Determine the total number of solutions (including complex and repeated) of the polynomial equation.

1. $4x^3 - 7x^2 + 5x - 9 = 0$

2. $8x^6 - 3x^4 - 11x^3 - 2x^2 + 4 = 0$

3. $x^5 + 2x^3 - 4x^2 + 7x = 12$

4. $-3x^4 + 2x^3 + 15x^2 - x + 1 = 8$

Decide whether the given x-value is a zero of the function.

5. $f(x) = x^4 + 2x^3 + 5x^2 + 8x + 4, x = -1$

6. $f(x) = x^4 - x^3 - 8x^2 + 2x + 12, x = 2$

7. $f(x) = x^3 + 4x^2 + x + 4, x = i$

8. $f(x) = 2x^3 - x^2 + 8x - 4, x = -2i$

Identify the factors of a polynomial function that has the given zeros.

9. $3, 1, 2$

10. $4, -1, -2, 0$

11. $-6, 2, 1, 1$

12. $-3, i, -i$

13. $4, -5, 2i, -2i$

14. $3, 2 + i, 2 - i$

Write a polynomial function of least degree that has real coefficients, the given zeros, and a leading coefficient of 1.

15. $-1, 2, 4$

16. $3, 1, 1$

17. $-3, 2, 0$

18. $-2, i, -i$

19. $0, 1, 3i, -3i$

20. $1, 2, i, -i$

21. $i, -i, 3i, -3i$

22. $2, 4 + i$

23. $-1, -1, 1 - 2i$

Find all of the zeros of the polynomial function.

24. $f(x) = x^3 + x^2 + x + 1$

25. $f(x) = 2x^3 + 3x^2 - 11x - 6$

26. $f(x) = x^3 + 4x^2 + 9x + 36$

27. $f(x) = 2x^4 + x^3 + x^2 + x - 1$

28. $f(x) = x^4 - 3x^2 - 4$

29. $f(x) = 3x^4 + 11x^3 + 8x^2 + 44x - 16$

30. *Preakness Stakes*   For 1990 through 1998, the value of a horse winning the Preakness Stakes can be modeled by

$$V = 2553x^3 - 25,200.56x^2 + 64,026.95x + 428,075.56$$

where x is the number of years since 1990. Use a graphing calculator to determine in what year the winnings were $488,150.

31. *NBA Standings*   There are seven teams in the Atlantic Division of the Eastern Conference of the NBA. During the 1997-98 season the winning percentage of the teams in this division can be modeled by

$$W = -0.0051x^3 + 0.063x^2 - 0.260x + 0.862$$

where x is the team's rank within the division. Orlando's winning percent was 0.500. Use a graphing calculator to estimate their ranking in the division.

NAME _____ DATE _____

Practice C

For use with pages 366–371

Determine the total number of solutions (including complex and repeated) of the polynomial equation.

1. $2x^2 + 3x = x^4 - 5x + 1$

2. $3 - 2x^2 + x^3 = 0$

3. $4 - 7x = x^2 - 3x^5$

Decide whether the given x-value is a zero of the function.

4. $f(x) = x^3 - 5x^2 - 4x + 6, x = 1 + i$

5. $f(x) = x^3 + 2x^2 - 3x - 10, x = 2 - i$

6. $f(x) = x^4 + 5x^3 + 5x^2 - 5x - 6, x = 2$

7. $f(x) = x^5 + 4x^4 + 10x^3 + 4x^2 + 9x + 36,$
$x = -3i$

Write a polynomial function of least degree that has real coefficients, the given zeros, and a leading coefficient of 2.

8. $-4, 0, 2, 4$

9. $2i, -2i, 5i, -5i$

10. $4 + i, 4 - i, -2$

11. $3 + 4i, 0$

12. $2 - i, 1, 2$

13. $5 - 2i, i, 0, 3$

Find all the zeros of the polynomial function.

14. $f(x) = x^3 - 17x^2 + 96x - 182$

15. $f(x) = x^4 - 4x^3 + 4x - 1$

16. $f(x) = x^4 + 12x^3 + 54x^2 + 108x + 81$

17. $f(x) = 2x^5 - 4x^4 - 2x^3 + 28x^2$

Find all the zeros of the polynomial function using the given hint.

18. $f(x) = x^4 + 2x^3 + 2x - 1$
  Hint: $-i$ is a zero

19. $f(x) = x^4 - 2x^3 + 14x^2 + 6x - 51$
  Hint: $1 + 4i$ is a zero

20. *College Tuition*   For 1990 through 1997 the enrollment of a college can be modeled by $E = -29.881t^2 + 190.833t + 4935$ where t is the number of years since 1990. For the same years, the cost of tuition at the college can be modeled by $T = 10.543t^3 - 118.826t^2 + 921.032t + 9978.758$ where t is the number of years since 1990. Write a model that represents the total tuition brought in by the college in a given year. In what year did the college take in $62,638,006 in tuition?

21. *Critical Thinking*   The graph of a polynomial of degree 5 has four distinct x-intercepts. Since the total number of solutions (including complex and repeated) must be 5, is the fifth solution a complex solution or a repeated solution? Explain your answer.

NAME _____ DATE _____

Reteaching with Practice

For use with pages 366–371

GOAL How to use the fundamental theorem of algebra to determine the number of zeros of a polynomial function and how to use technology to approximate the real zeros of a polynomial function

> ### VOCABULARY
>
> A factor that appears twice in a factored equation is called a **repeated solution.**

EXAMPLE 1 ## Finding the Number of Solutions or Zeros

Equations have solutions, whereas functions have zeros.

a. The equation $x^4 - 2x^3 + 10x^2 - 18x + 9 = 0$ has four solutions, since its degree is 4 : 1, 1, 3i and $-3i$. Notice 1 is a repeated solution.

b. The function $f(x) = x^3 + 7x^2 + 7x - 15$ has three zeros, since its degree is 3 : -3, 1, and -5.

Exercises for Example 1

Find the number of zeros of the polynomial function.

1. $f(x) = x^4 - x^3 - x^2 - x - 2$

2. $f(x) = x^3 - 8x^2 + 19x - 12$

3. $f(x) = x^3 - 2x^2 + 5x - 10$

4. $f(x) = x^4 + 3x^3 - 4x^2 - 12x$

EXAMPLE 2 ## Finding the Zeros of a Polynomial Function

Find all the zeros of $f(x) = x^4 - 2x^3 + 10x^2 - 18x + 9$.

SOLUTION

The possible rational zeros are $\pm1, \pm3,$ and $\pm9.$ Using synthetic division, you can determine that 1 is a repeated zero. The function in factored form is:

$$f(x) = (x - 1)(x - 1)(x^2 + 9)$$

Use the quadratic formula to factor the sum of two squares in order to factor completely.

$$x = \frac{-b \pm \sqrt{b^2 - 4ac}}{2a} = \frac{-0 \pm \sqrt{0^2 - 4(1)(9)}}{2(1)}$$

$$= \frac{\pm\sqrt{-36}}{2} = \frac{\pm 6i}{2} = \pm 3i$$

$$f(x) = (x - 1)(x - 1)(x + 3i)(x - 3i)$$

This factorization gives the four zeros of 1, 1, 3i, and $-3i$.

Reteaching with Practice

For use with pages 366–371

Exercises for Example 2

Find all the zeros of the polynomial function.

5. $f(x) = x^4 - x^3 - x^2 - x - 2$ **6.** $f(x) = x^3 - 8x^2 + 19x - 12$

7. $f(x) = x^3 - 2x^2 + 5x - 10$ **8.** $f(x) = x^4 + 3x^3 - 4x^2 - 12x$

EXAMPLE 3 · *Using Zeros to Write Polynomial Functions*

Write a polynomial function f of least degree that has real coefficients, a leading coefficient of 1, and the zeros -1, 5, and 6.

SOLUTION

Using the three zeros and the factor theorem, write $f(x)$ as a product of three factors.

$$f(x) = (x + 1)(x - 5)(x - 6) \qquad \text{Write } f(x) \text{ in factored form.}$$
$$= (x^2 - 5x + x - 5)(x - 6) \qquad \text{Multiply } (x + 1)(x - 5).$$
$$= (x^2 - 4x - 5)(x - 6) \qquad \text{Combine like terms.}$$
$$= x^3 - 4x^2 - 5x - 6x^2 + 24x + 30 \qquad \text{Multiply.}$$
$$= x^3 - 10x^2 + 19x + 30 \qquad \text{Combine like terms.}$$

Exercises for Example 3

Write a polynomial function of least degree that has real coefficients, the given zeros, and a leading coefficient of 1.

9. $2, -1, 5$ **10.** $4, -3, 6$ **11.** $-1, 1, 7$ **12.** $3, -3, -2$

EXAMPLE 4 *Approximating Real Zeros*

Approximate the real zeros of $f(x) = x^3 + 2x^2 - 5x + 1$.

SOLUTION

First, enter the function in a graphing calculator. Then use the *Zero* (or *Root)* feature as shown.

From this screen, you can see that one of the real zeros is about -3.51.

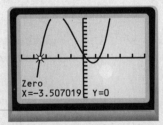

```
Zero
X=-3.507019  Y=0
```

Exercises for Example 4

Use a graphing calculator to graph the polynomial function. Then use the *Zero* (or *Root*) feature of the calculator to find the real zeros of the function.

13. $f(x) = x^3 - 6x + 2$ **14.** $f(x) = x^3 - x^2 + x - 1$

15. $f(x) = x^3 - 3x^2 + 4x + 3$ **16.** $f(x) = x^3 - 3x - 1$

NAME _____ DATE _____

Quick Catch-Up for Absent Students

For use with pages 366–372

The items checked below were covered in class on (date missed) _____

Lesson 6.7: Using the Fundamental Theorem of Algebra

_____ **Goal 1:** Use the fundamental theorem of algebra to determine the number of zeros of a polynomial function. (pp. 366, 367)

Material Covered:

_____ Activity: Investigating the Number of Solutions

_____ Example 1: Finding the Number of Solutions or Zeros

_____ Example 2: Finding the Zeros of a Polynomial Function

_____ Example 3: Using Zeros to Write Polynomial Functions

Vocabulary:

repeated solution, p. 366

_____ **Goal 2:** Use technology to approximate the real zeros of a polynomial function. (p. 368)

Material Covered:

_____ Example 4: Approximating Real Zeros

_____ Example 5: Approximating Real Zeros of a Real-Life Function

Activity 6.7: Solving Polynomial Equations (p. 372)

_____ **Goal:** Use a graphing calculator to solve polynomial equations.

_____ Student Help: Keystroke Help

_____ Other (specify) _____

_____

Homework and Additional Learning Support

_____ Textbook (specify) _pp. 369–371_____

_____

_____ Internet: Extra Examples at www.mcdougallittell.com

_____ *Reteaching with Practice* worksheet (specify exercises)_____

_____ *Personal Student Tutor* for Lesson 6.7

NAME _____ DATE _____

Real-Life Application:
When Will I Ever Use This?

For use with pages 366–371

High School Athletic Programs

Suppose you work for a sporting goods company that manufactures sports apparel for high school athletic programs. Your company is researching the recent trend of high school students that are participating in athletic programs.

Your company determines that the number of students S (in thousands) in the United States from 1984 through 1997 that participated in high school athletic programs can be modeled by the equation

$$S = 1.169t^3 - 28.77t^2 + 258.1t + 4402$$

where t represents the number of years since 1980.

1. Create a table of values for the model. According to your table, when did the number of students reach six million?

2. Sketch a graph of the model. Use a graphing calculator to verify your graph.

3. Write an equation that will determine the year the number of students reached 6 million. Then rewrite the equation by setting it equal to zero. Use a graphing calculator to approximate the real zero(s) of the rewritten equation.

4. Write an equation that will determine the year the number of students reached 7.1 million. Then rewrite the equation by setting it equal to zero. Use a graphing calculator to approximate the real zero(s) of the rewritten equation.

5. According to the model, during which year will the number of students that participate in high school athletic programs reach 8.7 million? 10.5 million?

6. In 2005, the goal of your company is to have at least 8% of the total number of students participating in high school athletic programs wearing its athletic apparel. Approximately how many high school students will be wearing the company's athletic apparel if the goal is reached?

Algebra 2
Chapter 6 Resource Book

Challenge: Skills and Applications

For use with pages 366–371

1. a. Let r_1, r_2, and r_3 be the zeros of the equation $x^3 + bx^2 + cx + d = 0$. Use the Factor Theorem to write the polynomial as a product of linear factors. Then express b in terms of r_1, r_2, and r_3.

b. Based on your answer to part (a), make a conjecture about the relationship between the coefficient of x^{n-1} in a polynomial equation of degree n and the zeros of the equation, given that the leading coefficient of the polynomial is 1.

2. a. Suppose the cubic equation $f(x) = ax^3 + bx^2 + cx + d = 0$ has three real zeros, two of which are equal (i.e. there is one repeated zero). Assuming $a > 0$, sketch a possible graph of $y = f(x)$. Use your graph to justify the assertion that $f(x) - k$ has 3 distinct zeros, for some real number k.

b. Using the graph that you drew in part (a), justify the assertion that $f(x) + m$ has only one real (nonrepeated) zero for some real number m.

3. Suppose r, s, A, B, and C are real numbers. Any polynomial function $f(x)$ has the property that if $f(r) = A$ and $f(s) = B$ and C is any point between A and B, then $f(t) = C$ for some number t between r and s.

a. Describe the property stated above geometrically in terms of the graph of $y = f(x)$.

b. Explain how the above property justifies the statement that if $f(x)$ is a polynomial function, $f(0) < 0$, $f(2) > 0$, then f has a zero between 0 and 2.

c. If $f(1) > 0$, what can you say about a zero of f? If $f(1) < 0$, what can you say?

d. Based on the idea of parts (a) and (b), describe an iterated method of finding an arbitrarily small interval on which $f(x)$ has a zero. (This is called the *Bisection Method*.)

4. In this exercise, you will prove that if $z = a + bi$ is a zero of a polynomial equation $f(x) = ax^n + bx^{n-1} + \cdots + jx + k = 0$ with real coefficients, then $z = a - bi$ is also a zero. (Recall that $\bar{x}$ is the conjugate of x.)

a. In a previous Challenge Set, you may have proved that, for any complex numbers z and w, $\bar{z} + \bar{w} = \overline{z + w}$ and $\bar{z} \cdot \bar{w} = \overline{zw}$. Use these facts to explain why $f(\bar{z}) = \overline{f(z)}$.

b. Use the fact that $\bar{0} = 0$, together with the fact stated in part (a), to prove that if z is a zero of $f(x) = 0$, then so is $\bar{z}$. (*Hint*: Take the conjugate of both sides of the equation $f(z) = 0$.)

Lesson Plan

1-day lesson (See *Pacing the Chapter,* TE pages 320C–320D)   **For use with pages 373–378**

GOALS **1. Analyze the graph of a polynomial function.**
2. Use the graph of a polynomial function to answer questions about real-life situations.

State/Local Objectives _____

_____

✓ Check the items you wish to use for this lesson.

STARTING OPTIONS

____ Homework Check: TE page 369; Answer Transparencies
____ Warm-Up or Daily Homework Quiz: TE pages 373 and 371, CRB page 104, or Transparencies

TEACHING OPTIONS

____ Motivating the Lesson: TE page 374
____ Lesson Opener (Visual Approach): CRB page 105 or Transparencies
____ Graphing Calculator Activity with Keystrokes: CRB pages 106–107
____ Examples 1–3: SE pages 373–375
____ Extra Examples: TE pages 374–375 or Transparencies; Internet
____ Closure Question: TE page 375
____ Guided Practice Exercises: SE page 376

APPLY/HOMEWORK

Homework Assignment

____ Basic   14–26 even, 30–32, 35–36, 42–43, 45–59 odd
____ Average   14–36 even, 37–39, 42–43, 45–59 odd
____ Advanced   14–36 even, 37–44, 45–59 odd

Reteaching the Lesson

____ Practice Masters: CRB pages 108–110 (Level A, Level B, Level C)
____ Reteaching with Practice: CRB pages 111–112 or Practice Workbook with Examples
____ Personal Student Tutor

Extending the Lesson

____ Applications (Interdisciplinary): CRB page 114
____ Challenge: SE page 378; CRB page 115 or Internet

ASSESSMENT OPTIONS

____ Checkpoint Exercises: TE pages 374–375 or Transparencies
____ Daily Homework Quiz (6.8): TE page 378, CRB page 118, or Transparencies
____ Standardized Test Practice: SE page 378; TE page 378; STP Workbook; Transparencies

Notes _____

_____

_____

TEACHER'S NAME _____ CLASS _____ ROOM _____ DATE _____

Lesson Plan for Block Scheduling

Half-day lesson (See *Pacing the Chapter,* TE pages 320C–320D)   For use with pages 373–378

GOALS   1. Analyze the graph of a polynomial function.
2. Use the graph of a polynomial function to answer questions about real-life situations.

State/Local Objectives _____

_____

_____

✓ **Check the items you wish to use for this lesson.**

CHAPTER PACING GUIDE	
Day	**Lesson**
1	6.1 (all); 6.2(all)
2	6.3 (all)
3	6.4 (all)
4	6.5 (all); 6.6(all)
5	6.7 (all); **6.8 (all)**
6	6.9(all); Review Ch. 6
7	Assess Ch. 6; 7.1 (all)

STARTING OPTIONS
____ Homework Check: TE page 369; Answer Transparencies
____ Warm-Up or Daily Homework Quiz: TE pages 373 and 371, CRB page 104, or Transparencies

TEACHING OPTIONS
____ Motivating the Lesson: TE page 374
____ Lesson Opener (Visual Approach): CRB page 105 or Transparencies
____ Graphing Calculator Activity with Keystrokes: CRB pages 106–107
____ Examples 1–3: SE pages 373–375
____ Extra Examples: TE pages 374–375 or Transparencies; Internet
____ Closure Question: TE page 375
____ Guided Practice Exercises: SE page 376

APPLY/HOMEWORK
Homework Assignment (See also the assignment for Lesson 6.7.)
____ Block Schedule: 14–36 even, 37–39, 42–43, 45–59 odd

Reteaching the Lesson
____ Practice Masters: CRB pages 108–110 (Level A, Level B, Level C)
____ Reteaching with Practice: CRB pages 111–112 or Practice Workbook with Examples
____ Personal Student Tutor

Extending the Lesson
____ Applications (Interdisciplinary): CRB page 114
____ Challenge: SE page 378; CRB page 115 or Internet

ASSESSMENT OPTIONS
____ Checkpoint Exercises: TE pages 374–375 or Transparencies
____ Daily Homework Quiz (6.8): TE page 378, CRB page 118, or Transparencies
____ Standardized Test Practice: SE page 378; TE page 378; STP Workbook; Transparencies

Notes _____

_____

_____

WARM-UP EXERCISES

For use before Lesson 6.8, pages 373–378

1. If 2 is a zero of a polynomial function $f(x)$, name a factor of $f(x)$.

2. If -2, 3, and 7 are x-intercepts of a polynomial function, what is the least possible degree of the function?

3. Does the graph of $y = 2x^2 - 5x + 4$ have a minimum or a maximum point?

4. Describe the end behavior of $f(x) = x(x - 6)(x + 2)$.

DAILY HOMEWORK QUIZ

For use after Lesson 6.7, pages 366–372

1. Decide whether $x = -3$ is a zero of $f(x) = 2x^3 - 3x^2 - 23x + 12$.

2. Find all the zeros of $f(x) = x^4 + 2x^3 - 7x^2 + 2x - 8$.

3. Write a polynomial function of least degree that has real coefficients, a leading coefficient of 1, and whose zeros are $-\sqrt{2}, -1, \sqrt{2}$.

4. Use a graphing calculator to find the real zeros of $f(x) = x^4 - 6x^3 + 8x^2 + 4x - 4$. Round to the nearest hundredth.

Lesson 6.8

Visual Approach Lesson Opener

For use with pages 373–378

A point on a graph of a polynomial function is a *turning point* if it is higher (or lower) than all of the nearby points on the graph. For example, the graph shows that the turning points for $y = x^3 - 3x^2 + 3$ are $(0, 3)$ and $(2, -1)$.

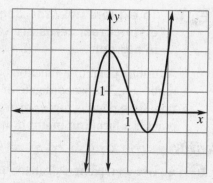

Find or approximate the *x*-intercepts and turning point or points of the graph.

1.

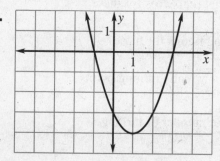

2.

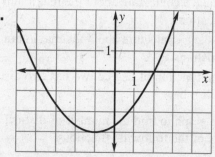

3.

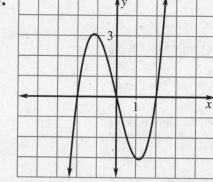

4.

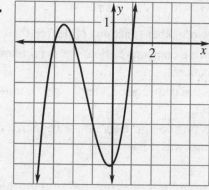

5.

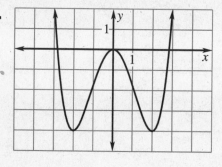

6.

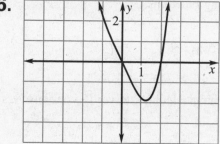

Graphing Calculator Activity

For use with pages 373–378

GOAL **To explore how the degree of a function and the number of distinct real zeros are related to the number of turning points in a graph**

A turning point of a function corresponds with a local maximum or a local minimum. The y-coordinate of a turning point is a **local maximum** of a function if the point is higher than all nearby points. The y-coordinate of a turning point is a **local minimum** if the point is lower than all nearby points.

Activity

1   **a.** Use a graphing calculator to graphically show that $f(x) = x^2$ has one distinct real zero and one turning point.

      b. Use a graphing calculator to graphically show that $f(x) = x^2 - 4x + 3$ has two distinct real zeros and one turning point.

2   **a.** Explore other functions of degree two.

      b. Make a conjecture about the maximum number of distinct real zeros for these functions.

      c. Make a conjecture about the maximum number of turning points for these functions

      d. If a function of degree two has two distinct real zeros, how many turning points will it have?

3   Use a graph to find the number of distinct real zeros and the number of turning points of the following functions.

$$f(x) = x^3 + x^2 - 5x + 3$$

$$f(x) = x^3 - 2x^2 - 5x + 6$$

4   **a.** Explore other functions of degree three.

      b. Make a conjecture about the maximum number of distinct real zeros for these functions.

      c. Make a conjecture about the maximum number of turning points for these functions.

      d. If a function of degree three has three distinct real zeros, how many turning points will it have?

Exercises

1. Determine the maximum number of distinct real zeros and the maximum number of turning points of the following functions.

a. $f(x) = x^4$

b. $f(x) = x^4 - 4x^2 + 4$

c. $f(x) = x^4 - x^3 - 7x^2 + x + 6$

2. Use a graphing calculator to verify your answers from Exercise 1.

NAME _____ DATE _____

Graphing Calculator Activity

For use with pages 373–378

TI-82

Step 1:

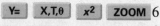

Step 3:

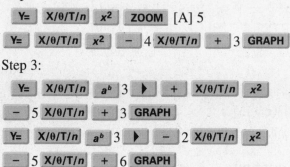

TI-83

Step 1:

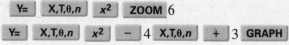

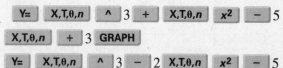

Step 3:

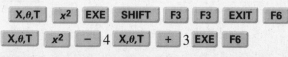

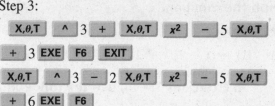

SHARP EL-9600c

Step 1:

Step 3:

CASIO CFX-9850GA PLUS

From the main menu, choose GRAPH.

Step 1:

Step 3:

NAME _______________________________ DATE __________

Practice A

For use with pages 373–378

Determine the lowest-degree polynomial that has the given graph.

1. 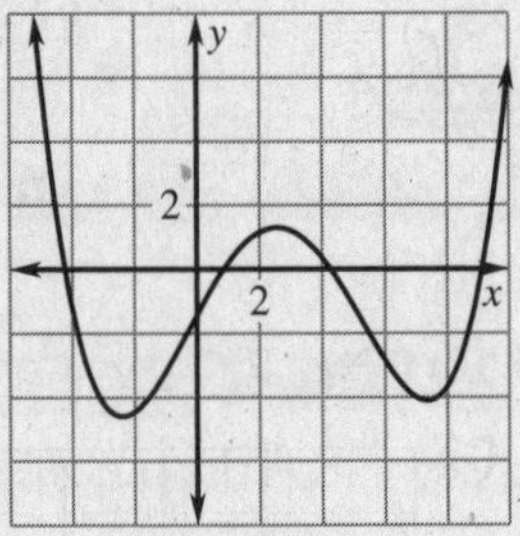

2. 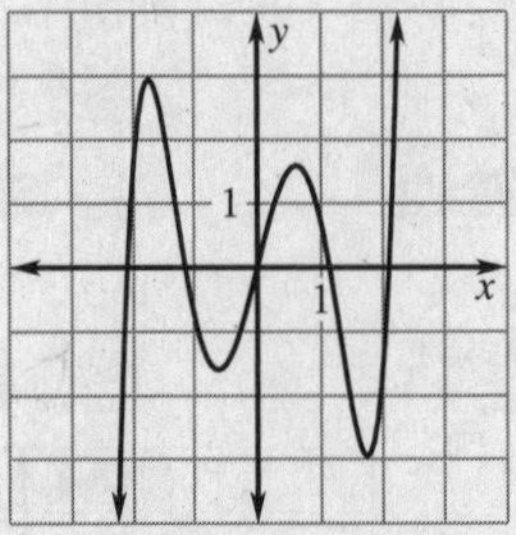

3. 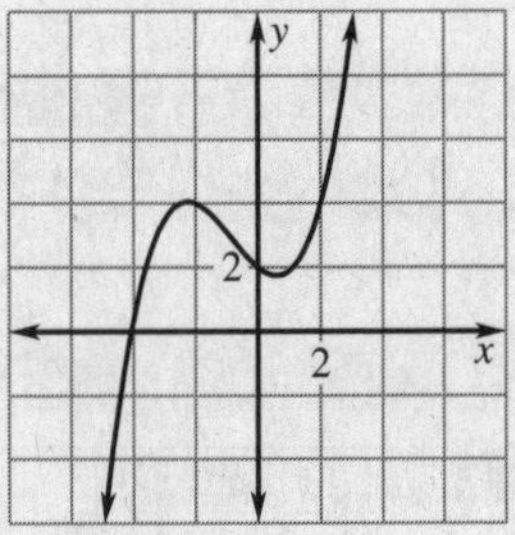

Estimate the coordinates of each turning point and state whether each corresponds to a local maximum or a local minimum.

4. 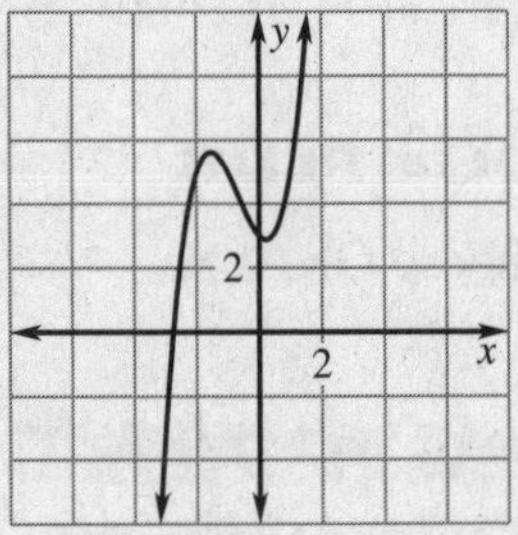

5. 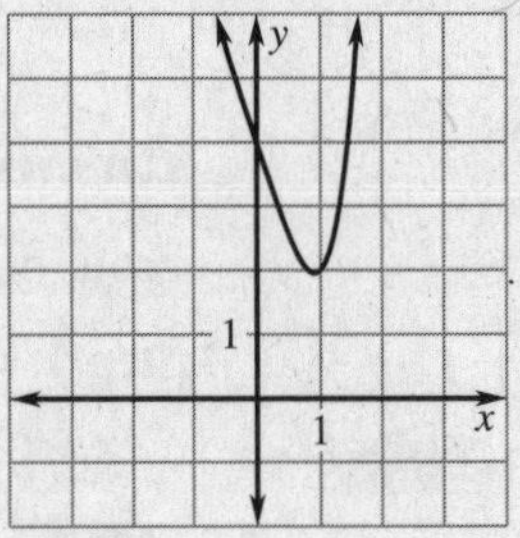

6. 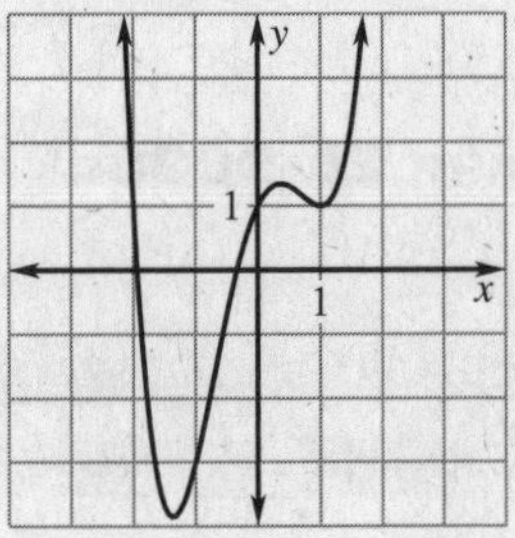

Graph the function.

7. $f(x) = (x - 2)(x - 4)$
8. $f(x) = (x - 1)(x + 3)$
9. $f(x) = (x + 2)(x - 4)$
10. $f(x) = (x - 1)(x + 2)(x + 3)$
11. $f(x) = (x - 3)(x - 1)(x + 1)$
12. $f(x) = (x - 2)(x + 2)^2$

Sales **In Exercises 13–15, use the following information.**

From 1990 to 1999, the annual sales S (in millions of dollars) of a certain company can be modeled by $S = 0.4t^3 - 4.5t^2 + 9.2t + 202$ where t is the number of years since 1990.

13. Use a graphing calculator to graph the polynomial function.
14. Approximate the year in which sales reached a low point.
15. If this polynomial function continues to model the sales of the company in the future, what can the expected sales be in 2000?

NAME _____ DATE _____

Practice B

For use with pages 373–378

State the maximum number of turns in the graph of the function.

1. $f(x) = x^4 + 2x^2 + 4$ 2. $f(x) = -3x^3 + x^2 - x + 5$ 3. $f(x) = 2x^6 + 1$

4. $f(x) = 4x^2 - 5x + 3$ 5. $f(x) = 3x^7 - 6x^2 + 7$ 6. $f(x) = 2x^9 - 8x^7 + 7x^5$

Match the graph with its function.

7. $f(x) = 2x^4 - 3x^2 - 2$ 8. $f(x) = 2x^6 - 6x^4 + 4x^2 - 2$ 9. $f(x) = -2x^4 + 3x^2 - 2$

A. 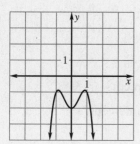   **B.** 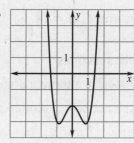   **C.**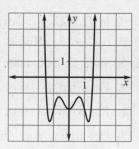

Determine the x-intercepts of the graph of the function.

10. $f(x) = (x + 3)(x - 2)(x - 5)$ 11. $f(x) = (x + 4)(x - 6)(x - 8)$ 12. $f(x) = (x + 3)^2(x - 2)$

13. $f(x) = (x + 5)(x + 1)(x - 7)$ 14. $f(x) = (x + 6)^3(x + 2)$ 15. $f(x) = (x - 8)^5$

Graph the function.

16. $f(x) = (x - 4)(x + 1)$ 17. $f(x) = (x - 3)(x + 4)(x + 1)$

18. $f(x) = (x - 3)^2(x + 2)$ 19. $f(x) = (x + 6)(x - 1)(x + 2)$

20. $f(x) = (x + 2)^2(x - 1)$ 21. $f(x) = (x + 1)^2(x - 1)(x - 4)$

22. $f(x) = x^3(x + 3)(x - 5)$ 23. $f(x) = (x - 1)(x^2 + x + 1)$

24. $f(x) = (x + 2)(x^2 + 2x + 2)$

25. **Olympic Platform Diving**   The polynomial function

     $P = 0.134t^3 - 5.775t^2 + 70.426t + 481.945$

models the number of points earned by the gold medal winner of the
platform diving event in the summer Olympics, where t is the number
of years since 1972. Graph the function and identify any turning points
on the interval $0 \le t \le 24$. What real-life meaning do these points have?
(*Hint:* The Olympics only take place every four years.)

26. **Livestock**   The polynomial function

     $C = 0.03t^4 + 3.53t^3 - 271.40t^2 + 3788.76t + 107,148.79$

models the number of cattle (in thousands) on farms from 1965 to 1998,
where t is the number of years since 1965. Graph the function and identify
any turning points on the interval $0 \le t \le 33$. What real-life meaning do
these points have?

NAME _____ DATE _____

Practice C

For use with pages 373–378

State the maximum number of turns in the graph of the function.

1. $f(x) = x^4 + 3x^3 - 2x + 5$ **2.** $f(x) = 4 - 2x^2 + 5x^3$ **3.** $f(x) = 2x - 3x^5 + 2x^2 - 5$

Graph the function.

4. $f(x) = (x + 3)^2(x - 1)$ **5.** $f(x) = (x - 2)^2(x - 5)^2$ **6.** $f(x) = (x + 1)^3(x - 3)^2$

7. $f(x) = (x + 2)^3(x - 1)^3$ **8.** $f(x) = (x + 2)^2(x - 3)(x + 1)$ **9.** $f(x) = (x - 10)^6$

Critical Thinking   **Consider the graphs $f(x) = (x + 1)^n$ where $n = 1, 2, 3, 4,$ and 5.**

10. What is the x-intercept for all of the functions?

11. For what values of n does the graph have a turning point at the x-intercept?

12. For what values of n does the graph not have a turning point at the x-intercept?

13. Generalize your findings in Exercises 11 and 12. Test your theory for $f(x) = (x + 1)^6$ and $g(x) = (x + 1)^7$.

Find all x-intercepts and identify the x-intercepts that are also locations of turning points for the graph of the function.

14. $f(x) = (x - 3)^2(x + 2)$ **15.** $f(x) = (x - 7)^3(x - 1)^3$

16. $f(x) = (x - 5)^6(x + 3)^2$ **17.** $f(x) = (x + 4)^8(x - 3)^5$

18. $f(x) = (x - 3)^5(x - 1)^2$ **19.** $f(x) = x(x - 2)(x + 4)^2$

20. $f(x) = (x + 3)(x - 2)(x - 5)$ **21.** $f(x) = (x + 3)^2(x - 2)(x - 3)^4$

22. $f(x) = (x + 8)^5(x - 1)^2(x - 4)^3$

23. *Consumer Economics*   The consumer price index of women's and girl's apparel from 1995 to 1998 can be modeled by

$$P = -1.02t^3 + 3.95t^2 - 4.53t + 131.5,$$

where t is the number of years since 1995. Graph the function and identify any turning points on the interval $0 \le t \le 3$. What real-life meaning do these points have?

NAME _____ DATE _____

Reteaching with Practice

For use with pages 373–378

GOAL **How to analyze the graph of a polynomial function and how to use the graph to answer questions about real-life situations**

VOCABULARY

The y-coordinate of a turning point is a **local maximum** of the function if the point is higher than all nearby points.

The y-coordinate of a turning point is a **local minimum** of the function if the point is lower than all nearby points.

EXAMPLE 1 *Using x-Intercepts to Graph a Polynomial Function*

Graph the function $f(x) = \frac{1}{3}(x + 1)^2(x - 3)^2$.

SOLUTION

First, plot the x-intercepts. Since $x + 1$ and $x - 3$ are factors of $f(x)$, -1 and 3 are the x-intercepts of the graph. Plot the points $(-1, 0)$ and $(3, 0)$. Second, plot points between and beyond the x-intercepts.

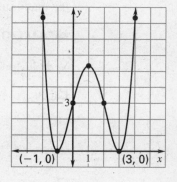

x	-3	-2	0	1	2	4	5
y	48	$8\frac{1}{3}$	3	$5\frac{1}{3}$	3	$8\frac{1}{3}$	48

Third, determine the end behavior of the graph. Since the degree is even and the leading coefficient is positive, $f(x) \to +\infty$ as $x \to -\infty$ and $f(x) \to +\infty$ as $x \to +\infty$.

Finally, draw a smooth curve that passes through the plotted points and has the appropriate end behavior.

Exercises for Example 1

Graph the function.

   1. $f(x) = (x - 1)^2(x + 2)$ **2.** $f(x) = (x - 1)(x + 2)^3$ **3.** $f(x) = (x + 3)^2(x + 5)^2$

EXAMPLE 2 *Finding Turning Points*

Graph $f(x) = x^5 - 3x^2 - 4x$ using a graphing calculator. Identify the x-intercepts and the local maximums and local minimums.

SOLUTION

Notice that the graph of the function has three x-intercepts and two turning points. Using the graphing calculator's *Zero* (or *Root*) feature, the x-intercepts are $x = -1, x = 0,$ and $x \approx 1.74$. Using the graphing calculator's *Maximum* and *Minimum* features, the approximate local minimum occurs at $(1.23, -6.64)$ and the local maximum occurs at $(-0.43, 1.15)$.

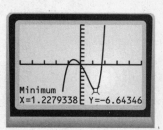

Algebra 2
Chapter 6 Resource Book

111

Reteaching with Practice

For use with pages 373–378

Exercises for Example 2

Use a graphing calculator to graph the polynomial function. Identify the *x*-intercepts and the points where the local maximums and local minimums occur.

4. $f(x) = -x^4 - 2x^3 + 3x^2 + 3x + 4$ **5.** $f(x) = x^3 - 3x^2 + 2x - 1$

6. $f(x) = x^4 - x^3 + x^2 - x - 1$ **7.** $f(x) = 2x^3 - 3x + 4$

EXAMPLE 3 *Maximizing a Polynomial Model*

You are designing a rain gutter made from a piece of sheet metal 1 foot by 5 feet. The gutter will be formed by turning up two sides. You want the rain gutter to have the greatest volume possible. How much should you turn up? What is the maximum volume?

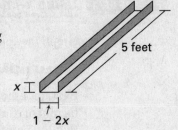

5 feet

x

$1 - 2x$

SOLUTION

Verbal Model: | Volume | = | Width | · | Length | · | Height |

Labels: Volume = V (cubic feet)

          Width = $1 - 2x$ (feet)

          Length = 5 (feet)

          Height = x (feet)

Algebraic Model: $V = (1 - 2x)(5)(x) = 5x - 10x^2$

Graph the polynomial using a graphing calculator. Then find the maximum volume using the *Maximum* feature. The maximum occurs at about $(0.25, 0.63)$. Therefore, you should turn up the sides 0.25 foot, or 3 inches. The maximum volume is about 0.63 cubic foot.

Exercises for Example 3

8. You are designing an open box to be made of a piece of cardboard that is 8 inches by 12 inches. The box will be formed by cutting and folding up the sides so that the flaps are square. You want the box to have the greatest volume possible. How many inches should you cut? What is the maximum volume?

9. What are the dimensions of the box in Exercise 8?

Algebra 2
Chapter 6 Resource Book

NAME _____ DATE _____

Quick Catch-Up for Absent Students

For use with pages 373–378

The items checked below were covered in class on (date missed) _____

Lesson 6.8: Analyze Graphs of Polynomial functions

_____ **Goal 1:** Analyze the graph of a polynomial function. (pp. 373, 374)

Material Covered:

_____ Example 1: Using x-Intercepts to Graph a Polynomial Function

_____ Example 2: Finding Turning Points

Vocabulary:

local maximum, p. 374   local minimum, p. 374

_____ **Goal 2:** Use the graph of a polynomial function to answer questions about real-life situations. (p. 375)

Material Covered:

_____ Example 3: Maximizing a Polynomial Model

_____ Other (specify) _____

_____

Homework and Additional Learning Support

_____ Textbook (specify) _pp. 376–378_____

_____

_____ Internet: Extra Examples at www.mcdougallittell.com

_____ *Reteaching with Practice* worksheet (specify exercises)_____

_____ *Personal Student Tutor* for Lesson 6.8

NAME _____ DATE _____

Interdisciplinary Application

Measuring Precipitation

SCIENCE   A rain gauge is the most common instrument used to measure the amount of rain that falls in a certain place over a specified length of time. The gauge consists of a cylinder with a narrow tube inside and a funnel on top. Rain falls into the funnel and down into the tube where it is measured. The mouth of the funnel has an area ten times that of the tube. So, an inch of rain that falls into the funnel would fill ten inches of the tube. A special ruler is used to measure the amount of rain in the tube.

Today's technology allows meteorologists to measure rainfall using radar. This electronic instrument sends out radio waves that are reflected by raindrops. These reflected waves indicate the amount and intensity of rainfall.

If an amount of rainfall is too small to be measured, it is called a *trace of rain.* Rainfall from a trace to 0.10 inch per hour is considered *light rain,* 0.11 to 0.30 inch per hour is a *moderate rain,* and anything greater than 0.30 inch per hour is classified as a *heavy rain.*

Average annual rainfall amounts vary throughout the world. Some factors that affect rainfall are latitude, bodies of water, mountains, air currents, and cities. Near the equator, the heat of the sun causes large amounts of moisture to evaporate into the air, while near the poles, it is so cold that the air cannot hold much moisture. The greatest amount of rainfall in the world occurs at Mount Waialeale in Hawaii where an average of 460 inches of rain falls each year. The least amount of rainfall in the world occurs at Arica, Chile, which receives an average of 0.03 inch each year.

In Exercises 1–3, use the following information.

The normal monthly precipitation P (in inches) in Juneau, Alaska for each month of the year can be approximated by the model

$$P = 0.00053t^5 - 0.02157t^4 + 0.2854t^3 - 1.412t^2 + 2.19t + 3.4$$

where $t = 1$ represents January.

1. Sketch a graph of the model for $1 \leq t \leq 12$.

2. From your graph, determine the month that has the greatest amount of precipitation and the least amount of precipitation.

3. What is the normal annual precipitation in Juneau, Alaska? Round your result to two decimal places.

Challenge: Skills and Applications

For use with pages 373–378

1. Let $f(x) = \frac{1}{4}(x - 3)(x + 1)^2$. Graph the function on a graphing calculator and note the x-intercepts. Now graph $f(x - 1)$, $f(x + 2)$, and $f(x - 4)$. How do these changes in the function affect the graph? State a general relationship between the graph of any polynomial function $y = f(x)$ and the graph of $y = f(x - a)$.

2. The *derivative* $f'(x)$ of a polynomial function $f(x)$ is the function whose value at x is the slope of the line tangent to the graph of $y = f(x)$ at the point $(x, f(x))$.

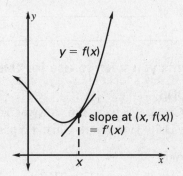

   a. Explain why the turning points of the graph of $y = f(x)$ are among the zeros of $f'(x) = 0$.

   b. Use the function $y = x^3$ to show that the derivative of a polynomial function may have a zero at a point that does not correspond to a turning point of the graph.

3. You can find the derivative (see Exercise 2) of a polynomial function as follows: In the expression for $f(x)$ replace every term ax^k by kax^{k-1}, and delete the constant term. For example, the derivative of $f(x) = x^3 + 5x^2 - 4x + 8$ is $f'(x) = 3x^2 + 10x - 4$.

   a. Find the derivative of $f(x) = \frac{1}{3}x^3 - x^2 - 3x + 4$, and use it to locate the turning points of the graph.

   b. Repeat part (a) for $f(x) = \frac{1}{4}x^4 - 2x^2 + 4$.

4. An *inflection point P* on the graph of a polynomial function is a point where the tangent line stops rotating one way—clockwise or counterclockwise—and starts rotating the other way, as you move from left to right. In the diagram, P is an inflection point of the graph.

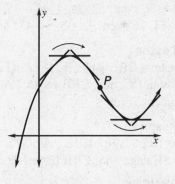

   a. By graphing the two functions in Exercise 3 on a graphing calculator, estimate the x-coordinates of all the inflection points of each graph. (There will be one for the cubic graph and two for the quartic graph.)

   b. The graph of a cubic equation $y = ax^3 + bx^2 + cx + d$ always has an inflection point at $x = \dfrac{-b}{3a}$. Find this value for the equation in Exercise 3(a).

TEACHER'S NAME _____ CLASS _____ ROOM _____ DATE _____

Lesson Plan

1-day lesson (See *Pacing the Chapter,* TE pages 320C–320D)   For use with pages 379–386

 GOALS
1. **Use finite differences to determine the degree of a polynomial function that will fit a set of data.**
2. **Use technology to find polynomial models for real-life data.**

State/Local Objectives _____

_____

✓ **Check the items you wish to use for this lesson.**

STARTING OPTIONS
____ Homework Check: TE page 376; Answer Transparencies
____ Warm-Up or Daily Homework Quiz: TE pages 380 and 378, CRB page 118, or Transparencies

TEACHING OPTIONS
____ Concept Activity: SE page 379; CRB page 119 (Activity Support Master)
____ Lesson Opener (Application): CRB page 120 or Transparencies
____ Examples 1–4: SE pages 380–382
____ Extra Examples: TE pages 381–382 or Transparencies; Internet
____ Closure Question: TE page 382
____ Guided Practice Exercises: SE page 383

APPLY/HOMEWORK
Homework Assignment
____ Basic   14–16, 18, 24, 32, 43, 47, 49–65 odd; Quiz 3: 1–21
____ Average   14–26 even, 32–34, 44–47, 49–65 odd; Quiz 3: 1–21
____ Advanced   14–26 even, 32–35, 44–48, 49–67 odd; Quiz 3: 1–21

Reteaching the Lesson
____ Practice Masters: CRB pages 121–123 (Level A, Level B, Level C)
____ Reteaching with Practice: CRB pages 124–125 or Practice Workbook with Examples
____ Personal Student Tutor

Extending the Lesson
____ Applications (Real-Life): CRB page 127
____ Challenge: SE page 385; CRB page 128 or Internet

ASSESSMENT OPTIONS
____ Checkpoint Exercises: TE pages 381–382 or Transparencies
____ Daily Homework Quiz (6.9): TE page 385 or Transparencies
____ Standardized Test Practice: SE page 385; TE page 385; STP Workbook; Transparencies
____ Quiz (6.7–6.9): SE page 386

Notes _____

_____

_____

TEACHER'S NAME _____ CLASS _____ ROOM _____ DATE _____

Lesson Plan for Block Scheduling

Half-day lesson (See *Pacing the Chapter,* TE pages 320C–320D)   For use with pages 379–386

 GOALS
1. **Use finite differences to determine the degree of a polynomial function that will fit a set of data.**
2. **Use technology to find polynomial models for real-life data.**

State/Local Objectives _____

_____

_____

✓ **Check the items you wish to use for this lesson.**

STARTING OPTIONS
_____ Homework Check: TE page 376; Answer Transparencies
_____ Warm-Up or Daily Homework Quiz: TE pages 380 and 378,
        CRB page 118, or Transparencies

TEACHING OPTIONS
_____ Concept Activity: SE page 379; CRB page 119 (Activity Support Master)
_____ Lesson Opener (Application): CRB page 120 or Transparencies
_____ Examples 1–4: SE pages 380–382
_____ Extra Examples: TE pages 381–382 or Transparencies; Internet
_____ Closure Question: TE page 382
_____ Guided Practice Exercises: SE page 383

APPLY/HOMEWORK
Homework Assignment
_____ Block Schedule: 14–26 even, 32–34, 44–47, 49–65 odd; Quiz 3: 1–21

Reteaching the Lesson
_____ Practice Masters: CRB pages 121–123 (Level A, Level B, Level C)
_____ Reteaching with Practice: CRB pages 124–125 or Practice Workbook with Examples
_____ Personal Student Tutor

Extending the Lesson
_____ Applications (Real Life): CRB page 127
_____ Challenge: SE page 385; CRB page 128 or Internet

ASSESSMENT OPTIONS
_____ Checkpoint Exercises: TE pages 381–382 or Transparencies
_____ Daily Homework Quiz (6.9): TE page 385 or Transparencies
_____ Standardized Test Practice: SE page 385; TE page 385; STP Workbook; Transparencies
_____ Quiz (6.7–6.9): SE page 386

Notes _____

_____

_____

CHAPTER PACING GUIDE	
Day	**Lesson**
1	6.1 (all); 6.2(all)
2	6.3 (all)
3	6.4 (all)
4	6.5 (all); 6.6(all)
5	6.7 (all); 6.8 (all)
6	**6.9(all)**; Review Ch. 6
7	Assess Ch. 6; 7.1 (all)

WARM-UP EXERCISES

For use before Lesson 6.9, pages 379–386

1. Write a cubic function that has real coefficients, zeros at $-2, 1, 3$, and a leading coefficient of 1.

2. Find $f(2)$ for $f(x) = 3x^3 - 4x^2 + x + 7$.

3. Solve the linear system.

$a + b + c = 9$

$4a + 2b + c = 5$

$9a + 3b + c = 7$

DAILY HOMEWORK QUIZ

For use after Lesson 6.8, pages 373–378

1. Graph the function $f(x) = \frac{1}{2}(x - 1)(x + 2)(x - 3)^2$.

2. Graph the function $f(x) = 2x^3 - 5x^2 + 1$. Identify the x-intercept and the points where the local maximum(s) and minimum(s) occur.

Activity Support Master

For use with page 379

$f(n) = n + 1$

n	1	2	3	4	5	6
f(n)	2	3	4	5	6	7
First-order differences		1	1	1	1	1

$g(n) = \frac{1}{2}(n + 1)(n + 2)$

n	1	2	3	4	5	6
g(n)						
First-order differences						
Second-order differences						
Third-order differences						

$h(n) = \frac{1}{6}(n + 1)(n + 2)(n + 3)$

n	1	2	3	4	5	6
h(n)						
First-order differences						
Second-order differences						
Third-order differences						

NAME _____ DATE _____

Application Lesson Opener

For use with pages 380–386

**In this lesson, you will learn how to use polynomial functions
to model real-life data.**

**The table gives data on personal savings as a percent of
disposable income for the years 1960–1995.**

Year	Savings
1960	6.4%
1965	7.6%
1970	8.4%
1975	9.0%
1980	8.2%
1985	6.9%
1990	5.0%
1995	4.8%

1. Make a scatter plot of the data.

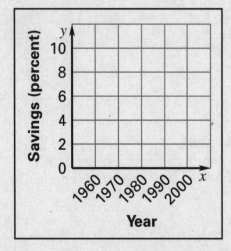

2. Suppose you want to model the data using a polynomial function.
Do you think it would be best to use a linear function, a quadratic
function, or a cubic function? Explain.

3. Sketch the polynomial function you think best fits the data.

Algebra 2
Chapter 6 Resource Book

NAME _____ DATE _____

Practice A

For use with pages 380–386

Match the cubic function with its graph.

  1. $f(x) = (x + 1)(x - 2)(x - 4)$ **2.** $f(x) = 2(x + 1)(x - 2)(x - 4)$ **3.** $f(x) = \frac{1}{2}(x + 1)(x - 2)(x - 4)$

A. 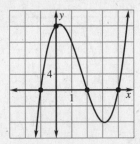   B. 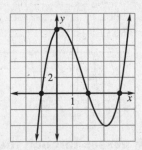   C.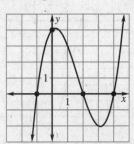

Write the cubic function whose graph is shown.

4. 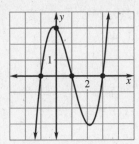   5. 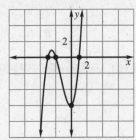   6.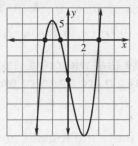

Write the cubic function whose graph passes through the given points.

  7. $(-3, 0), (-1, 0), (6, 0), (0, -18)$ **8.** $(-2, 0), (3, 0), (5, 0), (0, 30)$

  9. $(3, 0), (4, 0), (5, 0), (0, -60)$ **10.** $(-2, 0), (-1, 0), (6, 0), (0, -12)$

Show that the *n*th order differences for the given function of *n* are nonzero and constant.

  11. $f(x) = x^2 - 3x + 2$ **12.** $f(x) = x^3 + x^2 - x + 1$ **13.** $f(x) = 2x^3 + x^2 + 3x + 2$

Use a graphing calculator to find a cubic function for the data.

14.

x	0	1	2	3	4	5	6
y	11	15	20	16	14	16	18

15.

x	0	1	2	3	4	5	6
y	20	15	10	9	12	10	9

16.

x	0	1	2	3	4	5	6
y	53	40	30	24	23	10	-5

17.

x	0	1	2	3	4	5	6
y	2	5	7	7	9	11	20

NAME _____ DATE _____

Practice B
For use with pages 380–386

Write the cubic function whose graph is shown.

1.
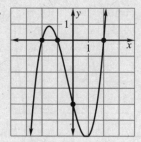

2.
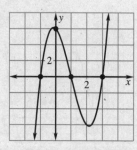

3.
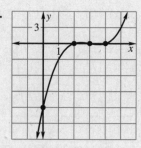

Write a cubic function whose graph passes through the given points.

4. $(-1, 0), (-3, 0), (2, 0), (0, 12)$ **5.** $(-4, 0), (-1, 0), (5, 0), (0, -10)$

6. $(-2, 0), (4, 0), (6, 0), (0, -48)$ **7.** $(-1, 0), (3, 0), (4, 0), (0, 24)$

8. $(0, 0), (1, 0), (8, 0), (2, 24)$ **9.** $(0, 0), (3, 0), (9, 0), (1, 4)$

Show that the *n*th order differences for the given function of degree *n* are nonzero and constant.

10. $f(x) = x^3 - 2x^2 + x - 1$ **11.** $f(x) = x^3 + 2x^2 - 5x - 1$ **12.** $f(x) = 2x^3 - 3x^2 + 4x - 6$

Use finite differences and a system of equations to find a polynomial function that fits the data.

13.

x	1	2	3	4	5	6
$f(x)$	5	19	49	101	181	295

14.

x	1	2	3	4	5	6
$f(x)$	-5	-6	-1	16	51	110

15.

x	1	2	3	4	5	6
$f(x)$	4	4	2	-2	-8	-16

16.

x	1	2	3	4	5	6
$f(x)$	-1	-8	-25	-58	-113	-196

17. *Average Miles Traveled*   The table shows the average miles traveled per vehicle (in thousands) from 1960 to 1996. Find a polynomial model for the data. Then predict the average number of miles traveled per vehicle in 2000.

t	1960	1965	1970	1975	1980	1985	1990	1995	1996
M	9.7	9.8	10.0	9.6	9.5	10.0	11.1	11.8	11.8

Practice C

For use with pages 380–386

Write the cubic function whose graph is shown.

1.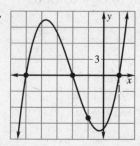

2.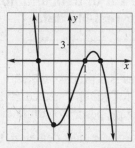

3.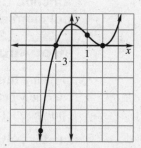

Write a cubic function whose graph passes through the given points.

4. $(-1, 0), (2, 0), (3, 0), (0, 9)$

5. $\left(-\frac{1}{2}, 0\right), (1, 0), (3, 0), \left(0, \frac{9}{4}\right)$

6. $\left(\frac{1}{2}, 0\right), \left(\frac{3}{2}, 0\right), (2, 0), (0, 18)$

7. $\left(-\frac{1}{3}, 0\right), \left(\frac{1}{4}, 0\right), (1, 0), (2, 49)$

8. $\left(-\frac{2}{3}, 0\right), \left(-\frac{1}{4}, 0\right), \left(\frac{1}{2}, 0\right), \left(1, \frac{25}{24}\right)$

9. $\left(-\frac{1}{3}, 0\right), \left(\frac{1}{3}, 0\right), \left(\frac{5}{6}, 0\right), (1, 16)$

Show that the *n*th order differences for the given function of degree *n* are nonzero and constant.

10. $f(x) = x^4 + 2x^3 - 3$

11. $f(x) = x^4 + x^3 - 3x^2 - 2x + 1$

12. $f(x) = x^5 - 4x^2 + 6$

13. $f(x) = x^4 - 8x$

Use finite differences and a system of equations to find a polynomial function that fits the data. You may want to use a calculator.

14.

x	1	2	3	4	5	6
$f(x)$	-16	-31	-54	-79	-100	-111

15.

x	1	2	3	4	5	6
$f(x)$	10	29	76	157	278	445

16. The table shows the U.S. population (in thousands) from 1990 to 1997.
Find a polynomial model for the data. Then estimate the U.S. population in 2000.

t	1990	1991	1992	1993
y	249,949	252,636	255,382	258,089

t	1994	1995	1996	1997
y	260,602	263,039	265,453	267,901

NAME _____ DATE _____

Reteaching with Practice

For use with pages 380–386

GOAL How to use finite differences to determine the degree of a polynomial function that will fit a set of data and how to use technology to find polynomial models for real-life data

VOCABULARY

Finding **finite differences** is a process which uses triangular numbers to decide whether y-values for equally spaced x-values can be modeled by a polynomial function. The properties of finite differences are listed below.

1. If a polynomial function $f(x)$ has degree n, then the nth-order differences of function values for equally spaced x-values are nonzero and constant.

2. Conversely, if the nth-order differences of equally spaced data are nonzero and constant, then the data can be represented by a polynomial function of degree n.

EXAMPLE 1 ### *Writing a Cubic Function*

Write the cubic function whose graph is shown.

SOLUTION

Begin by using the three x-intercepts to write the function in factored form:

$$f(x) = a(x + 1)(x - 2)(x - 3)$$

Then solve for a by substituting the coordinates of the point $(1, 8)$.

$$8 = a(1 + 1)(1 - 2)(1 - 3)$$

$$8 = 4a$$

$$2 = a$$

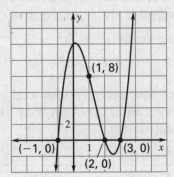

The cubic function is $f(x) = 2(x + 1)(x - 2)(x - 3)$.

Exercises for Example 1

Write the cubic function whose graph is shown.

1.

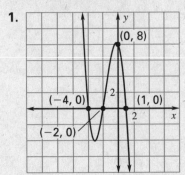

2.

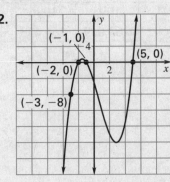

Reteaching with Practice

For use with pages 380–386

EXAMPLE 2 *Finding Finite Differences*

Show that the third-order differences for the function
$f(x) = x^3 - 2x^2 + x$ are nonzero and constant.

SOLUTION

Begin by evaluating the function for the first several values of x. For
example, $f(1) = 1^3 - 2(1)^2 + 1 = 0$

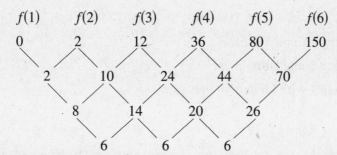

$f(1)$	$f(2)$	$f(3)$	$f(4)$	$f(5)$	$f(6)$	
0	2	12	36	80	150	Function values
	2	10	24	44	70	First-order differences
		8	14	20	26	Second-order differences
			6	6	6	Third-order differences

Notice that the third-order differences are nonzero and constant.

Exercises for Example 2

**Show that the *n*th-order differences for the given function of
degree *n* are nonzero and constant.**

3. $f(x) = 4x^2 - x + 5$ **4.** $f(x) = x^3 + x^2 - 3$

EXAMPLE 3 *Modeling with Cubic Regression*

Find a polynomial model that fits the data below.

x	1	2	3	4	5	6
$f(x)$	-1	5	33	95	203	369

SOLUTION

Enter the data in a graphing calculator. Use cubic regression to obtain a
model.

$$f(x) = 2x^3 - x^2 - 5x + 3$$

Exercises for Example 3

**Use a graphing calculator to find a polynomial function that
fits the data.**

5.

x	1	2	3	4	5	6
$f(x)$	8	10	10	8	4	-2

6.

x	1	2	3	4	5	6
$f(x)$	-12	-11	-2	21	64	133

NAME _____ DATE _____

Quick Catch-Up for Absent Students

For use with pages 379–386

The items checked below were covered in class on (date missed) _____

Activity 6.9: Exploring Finite Differences (p. 379)

_____ **Goal 1:** Find the relationship between the finite differences for a polynomial function and the function's degree.

Lesson 6.9: Modeling with Polynomial Functions

_____ **Goal 1:** Use finite differences to determine the degree of a polynomial function that will fit a set of data. (pp. 380, 381)

Material Covered:

_____ Example 1: Writing a Cubic Function

_____ Example 2: Finding Finite Differences

_____ Example 3: Modeling with Finite Differences

Vocabulary:

finite differences, p. 380

_____ **Goal 2:** Use technology to find polynomial models for real-life data. (p. 382)

Material Covered:

_____ Example 4: Modeling with Cubic Regression

_____ Other (specify) _____

_____

Homework and Additional Learning Support

_____ Textbook (specify) _pp. 383–386_____

_____

_____ Internet: Extra Examples at www.mcdougallittell.com

_____ *Reteaching with Practice* worksheet (specify exercises)_____

_____ *Personal Student Tutor* for Lesson 6.9

NAME _____ DATE _____

Real-Life Application: When Will I Ever Use This?

For use with pages 380–386

Business Ownership

Two main types of business ownership are partnerships and sole proprietorships. A partnership is an association formed by two or more people in a business enterprise. A sole proprietorship is a business that is owned and operated by one individual. Examples of partnerships can be found in law or medicine practices while sole proprietorships can include businesses such as beauty salons or small variety stores. The number of partnerships P (in thousands) and sole proprietorships S (in thousands) in the United States from 1980 through 1995 are shown in the table at the right where t represents the number of years since 1980.

t	P	S
0	1380	8,932
1	1461	9,585
2	1514	10,106
3	1542	10,704
4	1644	11,262
5	1714	11,929
6	1703	12,394
7	1648	13,091
8	1654	13,679
9	1635	14,298
10	1554	14,783
11	1515	15,181
12	1485	15,495
13	1468	15,848
14	1494	16,154
15	1581	16,424

1. Sketch two separate scatter plots for each set of data shown in the table.

2. Can either set of data be modeled by a linear equation? If so, find the linear model(s).

3. Use a graphing utility to find a cubic and quartic model for the number of partnerships P in the United States from 1980 through 1995. Which model better represents the data points?

4. Use your quartic model to predict the number of partnerships in 2005 and in 2010.

5. Do you think the quartic model could be used to predict the future number of partnerships? Explain.

6. Use a graphing calculator to find a cubic and quartic model for the number of sole proprietorships S in the United States from 1980 through 1995.

7. Could either of the models you found in Exercise 6 be used to predict the future number of sole proprietorships? Explain.

8. In general, what statement can be made about the number of partnerships and sole proprietorships in the United States?

Challenge: Skills and Applications

1. a. Let $f(x) = x^3 - 3x^2 + 2x + 5$. Find $f(x)$ for $x = -1, 0, 1, 2, 3, 4$, and 5, and use these values to compute the corresponding first differences for this function.

 b. Assume that the values you found in part (a) are function values for a function $g(x)$ for $x = -1, 0, 1, 2, 3$, and 4. Explain how you know that $g(x)$ must be a quadratic polynomial. Then find $g(x)$.

 c. Find the first differences for $g(x)$ for the values given. Find a function $h(x)$ whose values for $x = -1, 0, 1, 2$, and 3 are these differences.

2. a. Use the values $x = 1, 2$, and 3 to find the constant second difference for each of the functions $5x^2 - 2x + 3$, $-3x^2$, and $7x^2 + x - 1$. Express the constant (second) difference of each function in terms of the function's leading coefficient.

 b. Use the values $x = 1, 2, 3$, and 4 to find the constant third difference of each of the functions $-2x^3 + 3x^2 - x + 1$ and $5x^3 - 4x^2 - 3x + 2$. Express the constant (third) difference of each function in terms of the function's leading coefficient.

 c. Make a conjecture about the constant nth-order differences of a polynomial of degree n with leading coefficient a, using x-values that are spaced 1 unit apart. (*Hint:* The number $1 \cdot 2 \cdot 3 \cdot \cdots \cdot n$ is called "$n!$" (read "n factorial").)

3. a. Suppose you are given the constant third-order differences for a cubic polynomial for x-values spaced 1 unit apart. Explain why this information would not be sufficient to allow you to reconstruct the cubic polynomial.

 b. Let $f(x)$ be a cubic polynomial with zeros at $x = 1, 3$, and 4. Suppose that the (third) constant differences for $f(x)$ are 18 for x-values spaced 1 unit apart. Use this information and the result of Exercise 3 part (c) to find the cubic polynomial.

 c. Find $f(2)$ for the polynomial in part (b).

4. a. What will be the kth-order differences of a polynomial of degree n for $k > n$?

 b. Suppose $f(x) = g(x) + h(x)$, for some polynomials f, g, and h. Explain why any difference of $f(x)$ is the sum of the corresponding differences of $g(x)$ and $h(x)$.

 c. Use parts (a) and (b) to explain why terms of degree less than n do not matter when the nth difference is calculated.

 d. Use the result of part (c) to give a simplified method of finding the nth-order differences of a polynomial of degree n:

$$ax^n + bx^{n-1} + \ldots + rx + s.$$

Chapter Review Games and Activities

For use after Chapter 6

Solve the problems on the left. Match the answers with the column on the right.
Place the letters that are in front of the column on the right in the blanks below.
When they are on the line with the problem number you will answer the follow-
ing riddle. *What do you put on a sick pig?*

1. Simplify $\dfrac{(5x^{-3}y)^4}{15x^{10}y^3}$

2. Use synthetic substitution to evaluate the
   function for the given x value.

   $f(x) = x^3 - 6x^2 + 14x - 15$ for $x = 3$

3. Multiply $(x - 2)(x^2 + 4x - 8)$.

4. Solve for real solutions.

   $x^3 - 125 = 0$

5. Divide $(x^4 + 3x^3 + 2x^2 - x + 2) \div (x - 1)$.

6. Find all real zeros of the function
   $f(x) = x^3 - 6x^2 - x + 30$.

7. Solve $4x^3 + 8x^2 - x - 2 = 0$.

8. Factor $x^3 - 5x^2 + x + 15$.

(E) $x = 3, 5, -2$

(S) $x^3 + 2x^2 - x - 1$

(N) $x^3 + 2x^2 - 16x + 16$

(O) $\dfrac{125y}{3x^{22}}$

(B) $x = -5$

(I) $f(x) = 0$

(T) $(x - 3)(x^2 - 2x - 5)$

(M) $x^3 + 4x^2 + 6x + 5 + \dfrac{7}{x - 1}$

(A) $\dfrac{125x^2y}{3}$

(K) $x = 5$

(N) $x = \dfrac{1}{2}, -\dfrac{1}{2}, -2$

$\overline{}\ \overline{}\ \overline{}\ \overline{}$ -
(1) (2) (3) (4)

$\overline{}\ \overline{}\ \overline{}\ \overline{}$
(5) (6) (7) (8)

NAME _____ DATE _____

Chapter Test A

For use after Chapter 6

Simplify the expression.

1. $\dfrac{x^5}{x^6}$

2. $(2xy)^3$

3. $\dfrac{y^3}{y^{-3}}$

4. $\dfrac{25x^3y^2}{-5xy}$

Describe the end behavior of the graph of the polynomial function. Then evaluate for $x = -2, -1, 0, 1, 2$. Then graph the function.

5. $y = 3x^3 - 9x + 1$

x					
y					

6. $y = -x^3 + 4x$

x					
y					

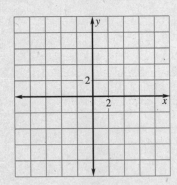

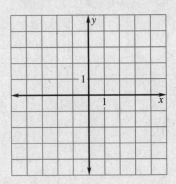

Perform the indicated operation.

7. $(x^2 - x + 1) + (x^2 - x + 1)$

8. $(2x + y)(2x - y)$

9. $(x + 1)(x^2 - x + 1)$

Factor the polynomial.

10. $25x^2 - 1$

11. $x^3 + 1$

12. $12x^4y^3 + 20x^2y^2 - 24x^2y$

Solve the equation.

13. $x^2 = 16$

14. $x^4 - 13x^2 + 36 = 0$

15. $x^3 + 4x^2 - x - 4 = 0$

Answers

1. _____

2. _____

3. _____

4. _____

5. Use grid at left.

6. Use grid at left.

7. _____

8. _____

9. _____

10. _____

11. _____

12. _____

13. _____

14. _____

15. _____

Divide. Use synthetic division if possible.

16. $(x^3 - 7x + 6) \div (x - 2)$ **17.** $(2x^3 + 6x^2 - 8) \div (x - 1)$

List all the possible rational zeros of *f* using the rational zero theorem. Then find all the zeros of the function.

18. $f(x) = x^2 + 4x + 3$ **19.** $f(x) = x^3 + x^2 - 10x + 8$

Write a polynomial function of least degree that has real coefficients, the given zeros, and a leading coefficient of 1.

20. $-4, -1, 3$ **21.** $4, 3$

22. Use technology to approximate the real zeros of
$f(x) = 0.25x^3 - x^2 + 2$.

23. Identify the x-intercepts, local maximum, and local minimum of the graph of $f(x) = \frac{1}{3}(x - 3)^2(x + 3)^2$. Then describe the behavior of the graph.

24. Show that the nth-order finite differences for the function $f(x) = x^2 - 4x + 4$ of degree n are nonzero and constant.

16. _____

17. _____

18. _____

19. _____

20. _____

21. _____

22. _____

23. _____

24. _____

Copyright © McDougal Littell Inc.
All rights reserved.

Algebra 2
Chapter 6 Resource Book

131

Review and Assess

NAME _____   DATE _____

Chapter Test B

For use after Chapter 6

Simplify the expression.

1. $\dfrac{x^3y^2}{x^4y}$ **2.** $(x^2y^3)^{-3}$ **3.** $\dfrac{x^4y^4}{x^{-4}y^{-4}}$ **4.** $\dfrac{xy}{1} \cdot (xy)^{-1}$

Describe the end behavior of the graph of the polynomial function. Then evaluate for $x = -2, -1, 0, 1, 2$. Then graph the function.

5. $y = -x^3$

x					
y					

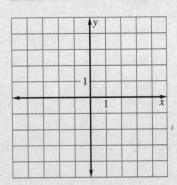

6. $y = 2x^3 + x^2 - 8x - 4$

x					
y					

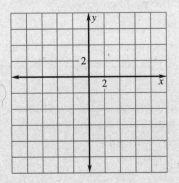

Perform the indicated operation.

7. $(3x^3 - x^2 + 4) - (2x^3 + x^2 + 2)$ **8.** $(x - 3y)(x - 4y)$

9. $(x + 1)(2x^2 - x + 1)$

Factor the polynomial.

10. $100x^2 - 9y^2$ **11.** $y^3 - 1$

12. $15x^3y^3 + 10x^2y^2 + 5xy$

Solve the equation.

13. $x^2 = 81$ **14.** $5x^3 = 30x - 25x^2$

15. $x(x + 5)(x - 4) = x^3$

Answers

1. _____

2. _____

3. _____

4. _____

5. Use grid at left.

6. Use grid at left.

7. _____

8. _____

9. _____

10. _____

11. _____

12. _____

13. _____

14. _____

15. _____

Divide. Use synthetic division if possible.

16. $(x^3 - 28x - 48) \div (x + 4)$

17. $(2x^3 + 11x^2 + 18x + 9) \div (x + 3)$

List all the possible rational zeros of f using the rational zero theorem. Then find all the zeros of the function.

18. $f(x) = x^2 - 6x + 5$ **19.** $f(x) = x^3 + x^2 - 10x + 8$

Write a polynomial function of least degree that has real coefficients, the given zeros, and a leading coefficient of 1.

20. $4, -5$ **21.** $-2, 2, 3$

22. Use technology to approximate the real zeros of
$f(x) = 0.35x^3 - 2x^2 + 8$.

23. Identify the x-intercepts, the local maximum, and local minimum of the graph of $f(x) = \frac{1}{4}(x - 2)^2(x + 2)^2$. Then describe the behavior of the graph.

24. Show that the nth-order finite differences for the function
$f(x) = -x^3 + 4x$ of degree n are nonzero and constant.

16. _____

17. _____

18. _____

19. _____

20. _____

21. _____

22. _____

23. _____

24. _____

NAME _____ DATE _____

Chapter Test C

For use after Chapter 6

Simplify the expression.

1. $(x^3y^2)^{-1}$

2. $\dfrac{x^3y^3}{x^{-2}y^{-2}}$

3. $\dfrac{1}{(xy)^{-3}}$

4. $\dfrac{x^2y^3}{y^{-4}} \cdot \dfrac{y^4}{x^{-2}y^{-3}}$

Describe the end behavior of the graph of the polynomial function. Then evaluate for $x = -2, -1, 0, 1, 2$. Then graph the function.

5. $y = x^3 - x^2 - 4x + 4$

x					
y					

6. $y = (x + 1)(x - 2)(x^2 - 3)$

x					
y					

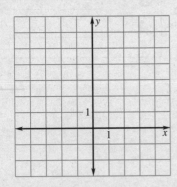

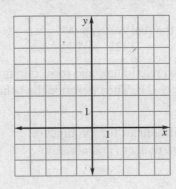

Perform the indicated operation.

7. $(4x^3 + 3x^2 - x + 2) - (5x^3 - 3x^2 + x - 4)$

8. $(xy + 4)(xy - 3)$

9. $(2x + y)(x^2 + xy + y^2)$

Factor the polynomial.

10. $16x^2 - 4y^2$

11. $8y^3 + 1$

12. $4c^3 + 8c^2d - 4cd^2 - 8d^3$

Solve the equation.

13. $2x^2 = 72$

14. $4y^3 + 48y^2 = 4y^4$

15. $(2x^2 + 3)^2 = 4x(x^3 + 6)$

Answers

1. _____

2. _____

3. _____

4. _____

5. Use grid at left.

6. Use grid at left.

7. _____

8. _____

9. _____

10. _____

11. _____

12. _____

13. _____

14. _____

15. _____

Review and Assess

Divide. Use synthetic division if possible.

16. $(x^3 - 2x^2 - 9) \div (x - 3)$

17. $(x^4 - 10x^2 + 2x + 3) \div (x - 3)$

List all the possible rational zeros of *f* using the rational zero theorem. Then find all the zeros of the function.

18. $f(x) = 2x^3 + x^2 + 2x + 1$　　**19.** $f(x) = x^3 + 2x^2 - 11x - 12$

Write a polynomial function of least degree that has real coefficients, the given zeros, and a leading coefficient of 1.

20. $-1, -2, -3$　　　　　　　　**21.** $3, -3, 2i, -2i$

22. Use technology to approximate the real zeros of
$f(x) = 0.2x^3 - 2x^2 + 6$.

23. Identify the *x*-intercepts, local maximum, and local minimum of the graph of $f(x) = \frac{1}{16}(x - 4)^2(x + 4)^2$. Then describe the behavior of the graph.

24. Show that the *n*th-order finite difference for the function
$f(x) = x^3 + 2x^2 - x - 2$ of degree *n* is nonzero and constant.

16. _____

17. _____

18. _____

19. _____

20. _____

21. _____

22. _____

23. _____

24. _____

Review and Assess

1. What is the value of 5^3?

  A 15   **B** $\frac{1}{15}$

  C 125   **D** 5

2. What is the value of $(-3)^3 \cdot \left(\frac{1}{2}\right)^2$?

  A $-\frac{27}{4}$ **B** $\frac{27}{4}$

  C $\frac{9}{4}$ **D** $-\frac{9}{4}$

3. Simplify $\dfrac{x^5 y^3}{x^2 y^{-3}}$.

  A x^3 **B** $x^7 y^6$

  C $\dfrac{1}{x^3 y^6}$ **D** $x^3 y^6$

4. What is the product of $(x + y)^3$?

  A $x^3 + y^3$

  B $x^3 + x^2 y^2 + y^3$

  C $x^3 + x^2 y + xy^2 + y^3$

  D $x^3 + 3x^2 y + 3xy^2 + y^3$

5. Factor $125x^3 - 64y^3$.

  A $(5x - 4y)(25x^2 - 20xy + 16y^2)$

  B $(5x - 4y)(25x^2 + 20xy + 16y^2)$

  C $(5x - 8y)^3$

  D $(5x + 4y)(25x^2 - 20xy + 16y^2)$

6. What are all the rational zeros of
$f(x) = x^3 - 7x - 6$?

  A none   **B** $1, 2, -3$

  C $-1, -2, 3$ **D** $-1, 2, -3$

7. How many zeros does the function
$f(x) = x^3 - 7x^2 + 7x + 15$ have?

  A none   **B** 3 rational

  C 2 real, 1 imaginary   **D** 3 imaginary

8. How many turning points does the function
$f(x) = -4x^3 + 15x^2 - 8x - 3$ have?

  A 2   **B** 1

  C 3   **D** 4

Quantitative Comparision In Exercises 9 and 10, choose the statement that is true about the given quantities.

  A The quantity in column A is greater.

  B The quantity in column B is greater.

  C The two quantities are equal.

  D The relationship cannot be determined from the given information.

9.

Column A	Column B
$-x^3$ when $x = 2$	$(x)^3$ when $x = -2$

  A   **B**   **C**   **D**

10.

Column A	Column B
The number of rational zeros in $f(x) = x^3 + 2x^2 - 11x - 12$	The number of rational zeros in $g(x) = x^3 - 9x^2 + 15x + 25$

  A   **B**   **C**   **D**

Alternative Assessment and Math Journal

For use after Chapter 6

JOURNAL 1. Examine the zeros of the following polynomial functions.

$$f(x) = x^4 + 5x^2 - 36$$

$$f(x) = x^4 - 12x^2 + 27$$

$$f(x) = x^3 - 3x^2 + 2$$

$$f(x) = x^3 - 3x^2 - 5x + 15$$

   a. What does it mean when two zeros of a polynomial function are complex conjugates? Which functions above, if any, have zeros that are complex conjugates?

   b. If you are asked to write a polynomial function of least degree with real coefficients and with zeros of 2 and $i\sqrt{7}$, what would be the degree of the polynomial? Explain.

MULTI-STEP
PROBLEM
2. In this exercise, refer to the division problem
$(4x^4 - 20x^3 + 23x^2 + 5x - 6) \div (x - 3)$.

   a. Find the quotient using long division.

   b. Find the quotient using synthetic division.

   c. Explain why you subtract in the process of long division, but add when using synthetic division.

   d. Under what conditions can you use synthetic division to determine the quotient? Give an example where synthetic division would not be a good option.

   e. What is the remainder for this problem? What information does this provide about $(x - 3)$?

   f. List the other possible zeros of the equation
   $f(x) = 4x^4 - 20x^3 + 23x^2 + 5x - 6$.

3. *Critical Thinking*   Use long division to divide $(2x - 3)$ into $4x^4 - x^2 - 2x + 1$. Then use synthetic division. What do you notice about your solutions? Explain why it is still possible to use synthetic division if the leading coefficient of the linear expression is not 1. Clearly indicate why one must be careful when using this method.

Alternative Assessment Rubric

For use after Chapter 6

JOURNAL
SOLUTION

1. **a–b.** Complete answers should address these points:

   a. • Explain that complex conjugates are zeros that come in pairs $a + bi$ and $a - bi$, where a and b are real numbers. The polynomial function $f(x) = x^4 + 5x^2 - 36$ has two zeros that are complex conjugates, $3i$, and $-3i$.

   b. • Explain that the polynomial must be of degree 3. If $i\sqrt{7}$ is a zero, then $-i\sqrt{7}$ must also be a zero. With three zeros, by the Fundamental Theorem of Algebra, the polynomial would be of degree 3.

MULTI-STEP
PROBLEM
SOLUTION

2. **a, b.** $4x^3 - 8x^2 - x + 2$

   c. • Answers will vary.

   d. • *Sample answer:* Synthetic division is used when dividing a polynomial by an expression of the form $x - k$. An example where synthetic division would not be used is $(4x^4 - 2x^2 + 3x - 1) \div (2x^3 + 4x - 1)$.

   e. • The remainder is zero. This means that $(x - 3)$ is a factor of the expression $4x^4 - 20x^3 + 23x^2 + 5x - 6$.

   f. • $\pm 6, \pm 3, \pm 2, \pm 1, \pm\frac{3}{2}, \pm\frac{3}{4}, \pm\frac{1}{2}, \pm\frac{1}{4}$

3. *Sample answer:* $2x^3 + 3x^2 + 4x + 5 + \dfrac{16}{2x - 3}$; It is possible to get this result using synthetic division. Factor a two out of the quotient and the divisor to get a leading coefficient of one for the divisor. Thus, the problem becomes

$$\left[2\left(2x^4 - \frac{1}{2}x^2 - x + \frac{1}{2}\right)\right] \div \left[2\left(x - \frac{3}{2}\right)\right].$$

After cancelling the twos, proceed with synthetic division as normal.

MULTI-STEP
PROBLEM
RUBRIC

4 Students complete all parts of the questions accurately. Explanations are clear and logical, showing an understanding of both synthetic and long division. Students show understanding of the rational zero theorem and remainder theorem. Students demonstrate thorough understanding by working the problem backwards.

3 Students complete the questions and explanations. Explanations may be somewhat vague, but show some understanding of long and synthetic division. Students demonstrate some understanding by trying to work the problem backwards.

2 Students complete questions and explanations. Explanations do not fit the questions. Students are unable to show understanding of both synthetic and long division. Examples are incomplete or inaccurate.

1 Students' work is very incomplete. Solutions or reasoning are incorrect. Examples are incomplete or inaccurate.

NAME _____ DATE _____

Project: Playing the Game

For use with Chapter 6

OBJECTIVE **Create a game that can be used by students to review polynomial functions.**

MATERIALS Old games, construction paper, poster board, index cards, marking pens, rulers, scissors, tape or glue

INVESTIGATION Games are fun, but they can also be a great learning tool. In this project, you will design a game based on the study of polynomial functions.

1. Your first task is to come up with a set of rules. Your game should require players to be familiar with as many concepts from Chapter 6 as possible.

2. If your game requires players to answer questions about polynomial functions, provide at least ten questions and answers to be used in your game. Otherwise, describe at least ten different situations that a player would encounter requiring knowledge of polynomial functions, and explain what the player should do in these situations.

3. Now comes the fun part! Decide on a name and design the box for your game. Create or provide all of the pieces or equipment to play the game. These may include a game board, question cards (including answers), dice, spinners, pencils, paper, and so on.

4. Test your game by playing it with others in your class. Make any adjustments that might be necessary.

PRESENT YOUR RESULTS Write a brief report explaining how you designed your game. If you worked in a group, include a paragraph describing your personal contribution to the completion of this project. If you needed to design another game, what would you do differently?

Review and Assess

Project: Teacher's Notes

For use with Chapter 6

GOALS • Write and answer various questions regarding polynomial functions.

MANAGING THE PROJECT

This project requires a lot of creativity, and the class presentations may take some time, so you may wish to have students work in groups.

Encourage groups to make collective decisions and to prepare the final report jointly. If necessary, you can break the report into parts and require each student to write the first draft of one part.

Questions you might want to ask to motivate students include: What are the most important skills you are learning in this chapter? What are some of the applications of polynomials? What are the rules of your favorite board games?

If students seem confused about how to create a game, you may wish to provide an example. The sample answer given in the back of the resource book may be used for this purpose.

RUBRIC **The following rubric can be used to assess student work.**

4 The rules and objects of the game are clear. Knowledge of polynomial functions is required to successfully play the game, and the questions provided by the student are of sufficient variety. The answers provided are correct. The game is well designed, fun to play, and presented in an attractive and appealing way.

3 The rules and objects of the game are clear. Knowledge of polynomial functions is required to successfully play the game, but the questions provided by the student may not have been chosen correctly or the student may have made a few errors in answering the questions. The game is reasonably well designed and appealing.

2 The game is playable, but work may be incomplete or reflect misunderstandings. For example, the rules may not have been explained adequately or many of the questions provided by the student may have been answered incorrectly. The report may indicate a limited grasp of certain ideas.

1 Portions of the game are missing. The student who created the game does not demonstrate an understanding of polynomial functions, or players are not required to understand polynomial functions to play the game.

Review and Assess

NAME _____ DATE _____

Cumulative Review

For use after Chapters 1–6

Evaluate the expression for the given values of the variables. (1.2)

1. $3x - 5$ when $x = -3$

2. $4x - (8x + 4)$ when $x = \frac{1}{4}$

3. $x^2 - 5x$ when $x = -2$

4. $x^3 - 4x^2 + x$ when $x = -2$

5. $x^3 + 2(x + 3)$ when $x = -3$

6. $-x^2 + 5x$ when $x = -2$

Solve the equation. (1.3)

7. $4x + 8 = 32$

8. $m - 15 = 3m + 4$

9. $4(3x - 5) = x + 3$

10. $\frac{3}{2}x + 4 = 2x + 1$

11. $\frac{1}{2}x + \frac{3}{4} = \frac{3}{2}x - \frac{29}{4}$

12. $\frac{2}{5}x + \frac{2}{3} = \frac{1}{10}x + \frac{11}{3}$

Solve the inequality. Then graph the solution. (1.6)

13. $2x + 5 \geq 9$

14. $5 - 2x < 15 + 3x$

15. $4x + 2 > 8$ or $4x + 2 < -10$

16. $3x - 7 > 8$ or $2x + 1 \leq -9$

17. $-5 < 3x + 1 < 10$

18. $-0.25 \leq 0.5x + 1 \leq 0.75$

Tell which line is steeper. (2.2)

19. Line 1: through $(-2, 5)$ and $(3, 7)$
Line 2: through $(0, 8)$ and $(-4, 3)$

20. Line 1: through $(0, -5)$ and $(3, -8)$
Line 2: through $(7, 1)$ and $(9, -10)$

21. Line 1: through $(4, 6)$ and $(5, 9)$
Line 2: through $(-3, 1)$ and $(5, 4)$

22. Line 1: through $(-5, 6)$ and $(-2, -3)$
Line 2: through $(-2, 8)$ and $(1, -9)$

Graph the equation using standard form. Label any intercepts. (2.3)

23. $3x + y = 8$

24. $2x - y = 7$

25. $4x + 3y = 12$

26. $5x - 4y = -2$

27. $y = 8$

28. $x = 6$

Write an equation of a line using the given information. (2.4)

29. The line passes through the point $(0, 5)$ and has a slope of 3.

30. The line passes through the point $(2, -4)$ and has a slope of $\frac{2}{5}$.

31. The line has a slope of $-\frac{3}{4}$ and a y-intercept of 5.

32. The line passes through the point $(2, 4)$ and is parallel to $x = 7$.

33. The line passes through the point $(2, -1)$ and is perpendicular to the line $y = \frac{2}{3}x + 5$.

34. The line passes through the point $(-4, 5)$ and is parallel to the line $y = -\frac{2}{3}x + 7$.

Evaluate the function for the given value of x. (2.7)

$f(x) = \begin{cases} 3x, & \text{if } x \leq 2 \\ x - 1, & \text{if } x > 2 \end{cases}$

35. $f(3)$

36. $f(2)$

37. $f(-1)$

38. $f(0)$

Tell how many solutions the linear system has. (3.1)

39. $4x + 2y = 8$
$8x + 4y = 16$

40. $3x + 2y = 6$
$-6x - 4y = 8$

41. $5x - 6y = 7$
$2x + 3y = 5$

Review and Assess

Cumulative Review

For use after Chapters 1–6

Solve the system using an algebraic method. (3.2)

42. $3x - 4y = 17$
$2x - y = 8$

43. $4x + y = 6$
$8x - 2y = -4$

44. $3x + 8y = 3$
$2x - 5y = 2$

45. $6x - 2y = -6$
$9x + 3y = 15$

46. $3x - \frac{1}{5}y = -7$
$\frac{1}{2}x + \frac{2}{5}y = 1$

47. $-2x + 5y = 0.5$
$4x - 3y = 1.1$

Evaluate the determinant of the matrix. (4.3)

48. $\begin{bmatrix} -4 & 3 \\ 4 & -1 \end{bmatrix}$

49. $\begin{bmatrix} 2 & 1 \\ 0 & -3 \end{bmatrix}$

50. $\begin{bmatrix} 3 & -2 \\ \frac{1}{2} & 2 \end{bmatrix}$

51. $\begin{bmatrix} 5 & 4 & 3 \\ 2 & -1 & 0 \\ 4 & -2 & 5 \end{bmatrix}$

52. $\begin{bmatrix} 0 & 1 & -3 \\ 5 & 1 & -2 \\ 6 & -1 & 7 \end{bmatrix}$

53. $\begin{bmatrix} -4 & 2 & 1 \\ 3 & 3 & 1 \\ -2 & 5 & 2 \end{bmatrix}$

Graph the quadratic function. (5.1)

54. $y = x^2 + 2x + 3$

55. $y = -x^2 + 4x + 5$

56. $y = 2x^2 + 8x - 3$

57. $y = (x - 1)^2 + 3$

58. $y = 2(x + 4)^2 - 2$

59. $y = -\frac{1}{2}(x + 5)^2 + 2$

Solve the quadratic equation. (5.2, 5.3)

60. $2x^2 + 5 = 11$

61. $\frac{1}{3}(x - 3)^2 = 2$

62. $-x^2 + 2 = -14$

63. $4x^2 + 12x + 9 = 0$

64. $x^2 - 9 = 0$

65. $3x^2 - 13x - 10 = 0$

Find the absolute value of the complex number. (5.4)

66. $3 - 4i$

67. $4 + 2i$

68. $1 - i$

69. $-1 + 5i$

70. $-3 - i$

71. $2 + \sqrt{7}i$

Use the quadratic formula to solve the equation. (5.6)

72. $2x^2 + x - 5 = 0$

73. $10x^2 - 9x + 1 = 0$

74. $x^2 - 8x + 9 = 0$

75. $3x^2 - x - 4 = 0$

76. $x^2 - 18 = 0$

77. $2x^2 + 4x = x^2 - 1$

Factor using any method. (6.4)

78. $9x^4 - 1$

79. $x^4 + 3x^2 + 2$

80. $6x^5 + 15x^3 + 6x$

81. $3x^4 - 81x$

82. $4x^4 + 37x^2 + 9$

83. $6x^3 + 9x^2 + 2x + 3$

Divide using synthetic division. (6.5)

84. $(2x^3 - 3x + 5) \div (x - 2)$

85. $(x^2 - 2x - 3) \div (x - 1)$

86. $(3x^2 - 4x + 6) \div (x - 4)$

87. $(x^4 - 2x^3 + x - 25) \div (x - 3)$

88. *Dimensions of a box* An open box with a volume of 32 in.³ is made from a square piece of metal by cutting 2-inch squares from each corner and then folding up the sides. Find the dimensions of the piece of metal required to make the box. **(5.5)**

ANSWERS

Chapter Support

Parent Guide

6.1: about 30 AU   **6.2:** $f(x) \to \infty$ as $x \to -\infty$ and $f(x) \to -\infty$ as $x \to \infty$.

6.3: $V = w^3 + 10w^2 + 21w$ **6.4:** $3, -3, -5$

6.5: $(x + 5)(3x - 4)(x - 2)$

6.6: 6 cm long by 4 cm wide by 8 cm high

6.7: $x^4 + x^3 - 8x^2 + 4x - 48$ **6.8:** x-intercepts are $-1, 1, -3, 3$; local maximum at $(0, 9)$; local minimums at $(2.24, -16)$ and $(-2.24, -16)$ **6.9:** $f(n) = 5n^3 - 5n^2 - 10n$

Prerequisite Skills Review

1. $-4x - 3$ **2.** $-3x^3 + 3x^2$

3. $-3x^4 + 2x^3 - 2x^2$ **4.** $-5x^2 - 12$

5.

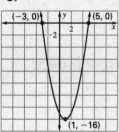

6.

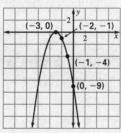

7.

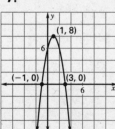

8.

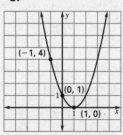

9. $y = x^2 - 8x + 13$ **10.** $y = -x^2 + 15x - 54$

11. $y = -x^2 + 10x - 24$ **12.** $y = 3x^2 - 3x - 6$

13. $-3, 8$ **14.** $-\frac{1}{2}, 7$ **15.** $4, 11$ **16.** $-10, 2$

Strategies for Reading Mathematics

1. a. The last term has x^{-2}, and -2 is not a whole number. **b.** first two terms are like terms and need to be combined. **c.** The terms are listed in order of increasing powers from left to right instead of decreasing powers. **d.** The constant term needs to be last.

2. a. $x^3 + x^2 - 3x + 12; 3; 1$

b. $-x^7 + 6x^3 - 7x^2 - 3x + 6; 7; -1$

3. a. $3x^2 - 2x - 7 = 0; 3; -2; -7$

b. $2x^2 - 4x + 6 = 0; 2; -4; 6$ (Also accept $-2x^2 + 4x - 6 = 0; -2; 4; -6$.)

c. $x^2 + 5x - 14 = 0; 1; 5; -14$

Lesson 6.1

Warm-ups

1. 16   **2.** 16   **3.** -16 **4.** 8   **5.** $\frac{1}{32}$

Daily Homework Quiz

1. $y = 2(x - 3)^2 - 4$ **2.** $y = 2(x - 1)(x + 1)$

3. $y = 2x^2 - 4x - 2$

Lesson Opener

1, 2. See tables below.

$	0.1	1
TL	45,000	450,000
TL (sci. not.)	$4.5 \times 10^{\boxed{4}}$	4.5×10^5

$	10	100
TL	4,500,000	45,000,000
TL (sci. not.)	$4.5 \times 10^{\boxed{6}}$	$4.5 \times 10^{\boxed{7}}$

TL	1	10
$	0.0000022	0.000022
$ (sci. not.)	2.2×10^{-6}	$2.2 \times 10^{\boxed{-5}}$

TL	100	1000
$	0.00022	0.0022
$ (sci. not.)	$2.2 \times 10^{\boxed{-4}}$	$2.2 \times 10^{\boxed{-3}}$

Lesson 6.1 *continued*

Answers

3. exponent increases by 1; exponent increases by 2; exponent increases by 3; exponent decreases by 1 **4.** $10^n \cdot 10^k = 10^{n+k}$

5. 4500TL; 1,000,000 TL

Practice A

1. 19,683 **2.** 256 **3.** 1024 **4.** 64 **5.** $\frac{1}{4}$

6. 5 **7.** $\frac{1}{216}$ **8.** $\frac{1}{243}$ **9.** $\frac{1}{128}$ **10.** -243

11. 25 **12.** $-\frac{1}{32}$ **13.** 25 **14.** $\frac{1}{343}$ **15.** 1

16. -32 **17.** $-\frac{1}{3}$ **18.** 15,625 **19.** 256

20. 729 **21.** $\frac{1}{81}$ **22.** $\frac{27}{8}$ **23.** $\frac{4}{25}$ **24.** $-\frac{1}{64}$

25. 1 **26.** 1 **27.** $\frac{1}{9}$ **28.** $\frac{1}{64}$ **29.** $\frac{16}{9}$ **30.** $\frac{81}{16}$

31. x^8 **32.** x^{12} **33.** x^{24} **34.** $27x^3$ **35.** $\frac{x^4}{16}$

36. x^5 **37.** $\frac{1}{x^6}$ **38.** $\frac{9}{x^2}$ **39.** $\frac{16}{x^2}$

40. 2.47×10^{10} mi^2 **41.** 1.88×10^8 mi^2

42. 5.59×10^7 mi^2

Practice B

1. 9 **2.** 15,625 **3.** $\frac{8}{27}$ **4.** $\frac{1}{64}$ **5.** 1 **6.** $\frac{1}{16}$

7. 9 **8.** 16 **9.** 1 **10.** x^5 **11.** $\frac{2}{y^2}$ **12.** $9x^2$

13. $\frac{y^3}{8}$ **14.** $256x^{12}$ **15.** $\frac{1}{y^2}$ **16.** $\frac{5x^3}{2y^2}$

17. $-\frac{y^5}{3x^2}$ **18.** $\frac{3}{2x^3}$ **19.** x^5 **20.** $\pi^3 x^4$

21. 7.02×10^1 people/mi^2 **22.** 1.08×10^5 mi/h

23. 580.52 computers/1000 people

Practice C

1. 16 **2.** 2187 **3.** 1 **4.** $\frac{27}{8}$ **5.** $\frac{81}{4096}$ **6.** 243

7. $\frac{x^3}{2y^2}$ **8.** $-\frac{4}{3y^2}$ **9.** 1 **10.** $-\frac{5y}{4x^2}$ **11.** $12x$

12. $\frac{4x^2}{y^4}$ **13.** $16x^5$ **14.** $\frac{y^{10}z^2}{4x^2}$ **15.** $\frac{1}{x^{48}}$ **16.** 2

17. 6 **18.** 4 **19.** 3 **20.** -5 **21.** -3

22. $\frac{4}{3}\pi\left(\frac{7}{4}\right)^3$ in.^{3} $= \frac{343\pi}{48}$ in.^{3}

23. $\frac{190 \cdot \frac{4}{3}\pi\left(\frac{7}{4}\right)^3}{(20)^3} = \frac{6517\pi}{38,400}$

24. about 53.32% **25.** 345,600 in.^{3}

Reteaching with Practice

1. 64 **2.** 4096 **3.** 6561 **4.** $\frac{8}{27}$ **5.** 25

6. 3 **7.** $32x^{10}$ **8.** $\frac{1}{y^5}$ **9.** $\frac{1}{9}$ **10.** $\frac{1}{x^6 y^{12}}$

11. $x^3 y^7$ **12.** 3.67×10^9 **13.** 7×10^7

Real-Life Application

1. a. 1.496×10^8 km **b.** 9.3×10^7 mi **2.** yes

3. 111,458 days **4.** $\frac{39.507}{0.387}$ **5.** 102 inches

6.

Planet	P	a
Mercury	2.41×10^{-1}	3.87×10^{-1}
Venus	6.15×10^{-1}	7.23×10^{-1}
Earth	1.00×10^0	1.00×10^0
Mars	1.881×10^0	1.523×10^0
Jupiter	1.1861×10^1	5.203×10^0
Saturn	2.9457×10^1	9.541×10^0
Uranus	8.4008×10^1	1.9190×10^1
Neptune	1.64784×10^2	3.0086×10^1
Pluto	2.4835×10^2	3.9507×10^1

Challenge: Skills and Applications

1. $2^m + 2^m = 2(2^m) = 2^{m+1}$

2. $2^n + 2^{n+1} = 2^n(1 + 2) = 3(2^n)$

3. a. 3; 5; 17; 257; 65,537

b. 4,294,967,297; 6,700,417

c. yes, no, no, yes, no, yes, yes, yes, yes

4. a. $M = \dfrac{rp(1 + r)^n}{(1 + r)^n - 1}$ **b.** \$1056.79

5. a. $1, -1$ **b.** $i^{2n} = (-1)^n$; $i^{2n+1} = (-1)^n i$

c. $i^{-2n} = (-1)^n$; $i^{-(2n+1)} = (-1)^n(-i) = (-1)^{n+1}i$

Lesson 6.2

Warm-ups

1. quadratic **2.** linear **3.** -13 **4.** 21

5. -14

Daily Homework Quiz

1. 16 **2.** 343 **3.** $\frac{y^9}{8x^7}$ **4.** $\frac{4x^3}{y^3}$ **5.** $\frac{3}{2}\pi x^3$

Lesson Opener
Allow 15 minutes.
1. no 2. yes 3. no 4. no 5. no 6. yes
7. yes 8. yes 9. no

Practice A
1. yes 2. yes 3. no 4. yes 5. no 6. yes
7. 5; 3 8. 7; -2 9. 4; 8 10. 2; $\frac{1}{3}$
11. 2; $\sqrt{5}$ 12. 9; 3 13. $f(x) = 2x^3 + 3x^2 - 5$
14. $f(x) = -5x^2 + 2x + 3$
15. $f(x) = 2x^3 + 5x^2 + 3x - 3$
16. $f(x) = -5x^2 + 3x + 14$
17. $f(x) = -5x^4 + 6x + 2$
18. $f(x) = -5x^3 + 7x^2 + x + 3$ 19. -1
20. 1 21. 34 22. -14 23. B 24. C
25. D 26. A
27. $C = 0.99x^2 + 14.93x + 75.32$ 28. 2; 0.99
29. ≈ 324

Practice B
1. yes; $f(x) = -2x^3 + 3x^2 + 4x$; 3; -2 2. no
3. no 4. yes; $f(x) = 2x^5 + 7x^2 - 3$; 5; 2
5. yes; $f(x) = -\frac{1}{6}x^2 + \frac{1}{3}x + \frac{2}{3}$; 2; $-\frac{1}{6}$
6. yes; $f(x) = \sqrt{5}\,x^4 + 2x^2 - x - 7$; 4; $\sqrt{5}$
7. -7 8. 3 9. 23 10. -30 11. 8
12. 52 13. 6 14. -84 15. 98 16. -86
17. -5 18. 74 19. 37 20. 0 21. -72
22. -6

23.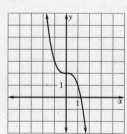
24.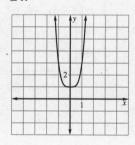

25.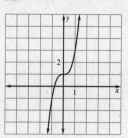
26.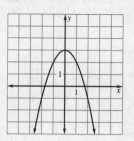

27.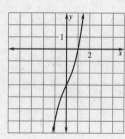
28.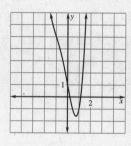

29.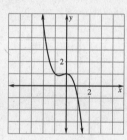
30.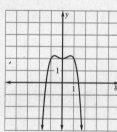

31.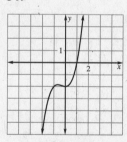

32. $1.02
33. $491,662.20

Practice C
1. 3 2. 15 3. 16 4. $\frac{35}{8}$ 5. -10
6. $13 + 5\sqrt{2}$ 7. 0 8. 45 9. $-\frac{61}{3}$ 10. $\frac{103}{10}$
11. $-\frac{9}{2}$ 12. $\frac{215}{27}$

Lesson 6.2 *continued*

13.

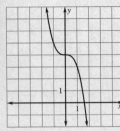

14.

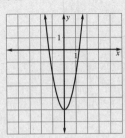

15.

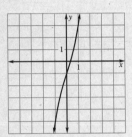

16.

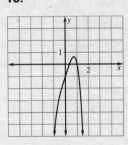

17.

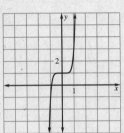

18.

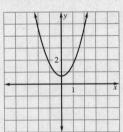

19.

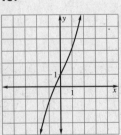

20.

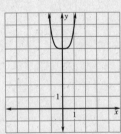

21.

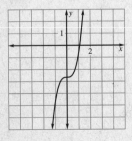

22. *Sample answer:* $f(x) = 3x^4 + 2x - 1$

23. *Sample answer:* $f(x) = -2x^3 + 3x^2 + x + 5$

24. 20.48 years   **25.** older

Reteaching with Practice

1. $f(x) = 5x^3 + x$; 3; cubic; 5

2. $f(x) = \pi x^4 - 4x + 6$; 4; quartic; π

3. not a polynomial function   **4.** 42   **5.** 75

6. 3   **7.** -2

8.

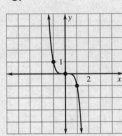

9.

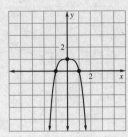

10.

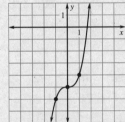

Interdisciplinary Application

1.

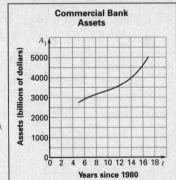

2. $f(x) \to +\infty$ as $x \to +\infty$; more than

3. about \$4246 billion

Lesson 6.2 *continued*

4.

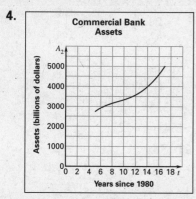

Commercial Bank Assets

Assets (billions of dollars) vs. Years since 1980

5. about \$4265 billion; the results are similar

6. A_1: \$23,851 billion; A_2: \$13,232 billion; the results from each model differ greatly

7. A_1 better represents the data because the model A_2 will eventually decrease rapidly and become negative.

Challenge: Skills and Applications

1. $4 + 2i$ **2.** $5i$ **3.** i **4.** $1 + i$

5. a. $\dfrac{a + \dfrac{b}{x} + \dfrac{c}{x^2} + \dots}{r + \dfrac{s}{x} + \dfrac{t}{x^2} + \dots}$

Dividing the fraction by $\dfrac{x^n}{x^n} = 1$ does not change the value of the fraction.

b. All the terms, except the first, in both numerator and denominator approach 0. Therefore, the whole expression approaches $\dfrac{a}{r}$. Yes.

6. a. They are the digits of the number. Since $10^j - 1 = 999\dots9$, which is divisible by 9, $10^j - 1 = 9v$, for some integer v. Therefore, $10^j = 9v + 1$.

b. $a \cdot 10^k + b \cdot 10^{k-1} + \dots + r \cdot 10 + s$
$= a(9v + 1) + b(9w + 1) + \dots + r(9z + 1) + s = 9(v + w + \dots + z) + (a + b + \dots + r + s) = $ a multiple of 9 + the sum of the digits. Since the multiple of 9 leaves no remainder when divided by 9, n and $(a + b + \dots + r + s)$ leave the same remainder when divided by 9. **c.** n will be divisible by 9 $\leftrightarrow$ it leaves a remainder of 0 when divided by 9 $\leftrightarrow$ the sum of the digits leaves a remainder of 0 $\leftrightarrow$ the sum of the digits is divisible by 9.

7. a.

k	a	b	c	d
		ak	$ak^2 + bk$	$ak^3 + bk^2 + ck$
	a	$ak + b$	$ak^2 + bk + c$	$ak^3 + bk^2 + ck + d$

b. From part (a), you have $ak^3 + bk^2 + ck + d = 0$; $d = -(ak^3 + bk^2 + ck)$, where $k = 2$. Therefore, the quantity in parentheses is an even integer, so d is even.

Lesson 6.3

Warm-ups

1. $2x - 10$ **2.** $-4x^2 + 20x - 4$ **3.** $6x^2$
4. a^7 **5.** $-5m^2$

Daily Homework Quiz

1. yes; $f(x) = -x^3 + 4x^2 + \frac{3}{5}x + 7$; 3; cubic; -1 **2.** 18 **3.** b

Lesson Opener

Allow 20 minutes.

1. C **2.** H **3.** E **4.** S **5.** S **6.** C **7.** H
8. A **9.** M **10.** P **11.** I **12.** O **13.** N
14. S

Graphing Calculator Activity

1. a. $3x^2 + 6x$ **b.** $x^2 - 3x$ **c.** $2x^2 + 4x + 12$
d. $2x^3 + 5x^2 + 16x + 2$ **e.** $x^2 + 3x - 10$
f. $6x^2 - 13x - 5$

Practice A

1. $3x^2 + 5x + 6$ **2.** $2x^3 - 5x^2 + x + 4$
3. $4x^4 + 3x^3 - 2x^2 + 4x - 8$ **4.** $4x^2 + 5x$
5. $5x^5 - 3x^4 + 2x^3 + x^2 - 3x - 8$
6. $-2x^2 - 3x + 8$ **7.** $-4x^3 + 6x^2 - 2x - 2$
8. $x - 4$ **9.** $2x^3 - 2x^2 + 2x - 2$
10. $-x + 2$ **11.** $-x + 3$ **12.** $-x^2 + 2x - 7$
13. $2x^2 + 7x - 3$ **14.** $-x^3 - 5x^2 + 8x - 5$
15. $2x^5 - 5x^2 + 9$ **16.** $-x^{12} + 5x^8 - 5x - 4$
17. $2x^3 - 4x^2 - 15x - 4$
18. $-x^3 + 2x^2 + 10x + 7$ **19.** $14x^2 + 9x - 18$
20. $x^2 + x - 12$ **21.** $x^2 - 8x + 12$
22. $x^2 + 5x + 6$ **23.** $2x^2 + 5x + 3$

Lesson 6.3 *continued*

24. $2x^2 - 11x + 5$ **25.** $3x^2 - x - 2$

26. $x^2 - 16$ **27.** $x^2 - 49$ **28.** $x^2 + 6x + 9$

29. $x^2 + 12x + 36$ **30.** $x^2 - 16x + 64$

31. $x^2 - 8x + 16$ **32.** $x^2 + 5x$

33. $2x^2 + 5x + 3$ **34.** $x^2 + 10x + 25$

35. $M = 3052.04t + 515,887.88$

Practice B

1. $2x^2 + 2x - 2$ **2.** $3x^3 - 3x^2 - x - 4$

3. $2x^2 + 6x - 4$ **4.** $-4x^2 - 2x + 4$

5. $7x^3 - 3x^2 - 2x + 1$

6. $2x^3 - 4x^2 + 3x + 1$ **7.** $3x^4 - 2x^2 - x + 5$

8. $2x^5 - x^3 + x^2 - 7x + 4$

9. $2x^5 + 3x^4 - x^2 - 5x + 4$

10. $-x^3 + 3x^2 - 8x + 5$ **11.** 10 **12.** $8x^2$

13. $3x^2 - x$ **14.** $2x^3 + 6x^2$ **15.** $3x^3 - x^2 + 5x$

16. $x^2 + 3x - 10$ **17.** $x^2 + 4x + 3$

18. $x^2 - 5x + 4$ **19.** $2x^2 + 11x + 5$

20. $3x^2 - 11x - 4$ **21.** $2x^2 - 5x + 3$

22. $6x^2 + 13x - 5$ **23.** $8x^2 - 10x - 3$

24. $15x^2 - 17x + 4$ **25.** $x^3 + 2x^2 - 1$

26. $x^3 - 7x - 6$ **27.** $x^3 - 3x^2 - 10x$

28. $x^2 - 81$ **29.** $4x^2 - 25$

30. $x^2 + 20x + 100$ **31.** $16x^2 + 24x + 9$

32. $x^2 - 24x + 144$ **33.** $9x^2 - 48x + 64$

34. $x^2 + 42x + 360$

35. $v = -16.2t^3 + 183t^2 + 1352.5t + 11504.1$

Practice C

1. $-x^3 - 3x^2 - 5x + 4$

2. $-x^3 + 3x^2 - 2x - 3$ **3.** $-5x^2 + 5x - 6$

4. $3x^4 + 4x^2 - 6x + 5$ **5.** $\frac{3}{2}x^2 + \frac{11}{3}x - 2$

6. $x^2 - 5x - \frac{2}{3}$ **7.** $-\frac{3}{10}x^3 + 3x^2 - 2x + 1$

8. $-\frac{3}{8}x^2 + \frac{1}{6}x - 5$ **9.** $6x^2 + 25x + 25$

10. $5x^2 - 32x - 21$ **11.** $24x^2 - 35x + 4$

12. $x^3 - x^2 - x + 1$ **13.** $2x^3 + 5x^2 - x - 4$

14. $2x^3 + 5x^2 + x - 2$ **15.** $2x^3 - x^2 - 7x - 3$

16. $-x^4 - 6x^3 + 5x^2 + 18x - 6$

17. $x^5 - 3x^4 - x^3 + 6x^2 - 12x + 9$

18. $x^6 + x^5 - 2x^4 - 6x^3 + x^2 + 10x - 5$

19. $2x^7 + 6x^6 - 3x^5 + 3x^4 + x$ **20.** $36x^2 - 25$

21. $\frac{1}{4}x^2 - 49$ **22.** $\frac{16}{9}x^2 + \frac{40}{3}x + 25$

23. $25x^2 - 20x + 4$ **24.** $\frac{1}{9}x^2 - \frac{4}{9}x + \frac{4}{9}$

25. $x^3 + 6x^2 + 12x + 8$

26. $x^3 - 9x^2 + 27x - 27$

27. $8x^3 + 12x^2 + 6x + 1$

28. $27x^3 - 135x^2 + 225x - 125$

29. $8x^3 + 36x^2y + 54xy^2 + 27y^3$

30. $16x^2 - 9y^2$ **31.** $36x^2 + 12xy + y^2$

32. $x^2 - 8xy + 16y^2$ **33.** $x^3 + 4x^2 + x - 6$

34. $x^3 - 4x^2 - 11x + 30$

35. $2x^3 + 9x^2 + 10x + 3$

36. $4x^3 - 20x^2 + 31x - 15$

37. $I = 814,536.25t^3 + 4,028,984.354t^2 +$
$17,858,746.41t + 560,699,692.4$

Reteaching with Practice

1. $9x^2 + 5x - 7$ **2.** $-2x^3 - 1$

3. $4x^2 + 8x - 4$ **4.** $2x + 9$ **5.** $x^2 - 4x - 2$

6. $4x^3 + 13x + 4$ **7.** $-10x^2 + x + 8$

8. $6x^2 - 2x + 10$ **9.** $3x^3 + 3x^2 - 6x$

10. $-2x + 2x^2 + 2x^3$ **11.** $x^3 - 8$

12. $2x^3 - x^2 - 7x - 3$ **13.** $x^2 - 25$

14. $4x^2 - 49$ **15.** $x^2 + 12x + 36$

16. $x^2 - 6x + 9$ **17.** $x^3 - 3x^2 + 3x - 1$

18. $x^3 + 6x^2 + 12x + 8$

Real-Life Application

1. Triangular-base pyramid:

Number of layers, n	2	3	4	5	6	7
Total number of cans in pyramid, C	4	10	20	35	56	84

Number of layers, n	8	9	10	11	12
Total number of cans in pyramid, C	120	165	220	286	364

Lesson 6.3 *continued*

Square-base pyramid:

Number of layers, n	2	3	4	5	6	7
Total number of cans in pyramid, C	5	14	30	54	90	139

Number of layers, n	8	9	10	11	12
Total number of cans in pyramid, C	203	284	384	504	648

2.

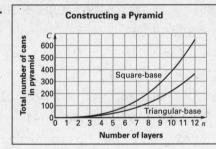

The number of cans used to construct the square-base pyramid increases more rapidly than the number of cans in the triangular-base pyramid.

3. $\frac{1}{6}n^3 + n^2 + \frac{11}{6}n + 1$

4. Find the difference between $\frac{1}{6}n^3 + n^2 + \frac{11}{6}n + 1$ and $\frac{1}{6}n^3 + \frac{1}{2}n^2 + \frac{1}{3}n$; $\frac{1}{2}n^2 + \frac{3}{2}n + 1$. **5.** $\frac{1}{3}n^3 + \frac{3}{2}n^2 + 2n + 1$

6. Find the difference between $\frac{1}{3}n^3 + \frac{3}{2}n^2 + 2n + 1$ and $\frac{1}{3}n^3 + \frac{1}{2}n^2 + \frac{1}{6}$; $n^2 + 2n + \frac{5}{6}$. **7.** 1327 cans **8.** 354 cans

Challenge: Skills and Applications

1. $p^4 - q^4$ **2.** $x^6 - y^6$

3. $(a^2 + 2ab + b^2) - c^2$

4. $u^4 - v^2 + 2vw - w^2$ **5. a.** $x^3 + y^3$

b. $x^5 - y^5$

c. $(x + y)(x^{2n} - x^{2n-1}y + x^{2n-2}y^2 - \ldots + y^{2n}) = x^{2n+1} + y^{2n+1}$

6. a. $a^4 + 4a^3b + 6a^2b^2 + 4ab^3 + b^4$

b. $x^8 + 8x^6 + 24x^4 + 32x^2 + 16$

c. $a^5 + (1 + 4)a^4b + (4 + 6)a^3b^2 + (6 + 4)a^2b^3 + (4 + 1)ab^4 + b^5$

d. Each coefficient of $(a + b)^n$ is the sum of 2 consecutive coefficients of $(a + b)^{n-1}$.

7. When you expand, you get $(n^k + rn^{k-1} + \ldots) - n^k$, so the terms of degree k have a sum of 0, and $r \neq 0$.

8. a. $f(0) = d$, but on the other hand $f(0) = 0^2 = 0$, so $d = 0$. **b.** $f(n + 1)$ contains the same terms as $f(n)$ but with the addition of one more term, $(n + 1)^2$, so this term is left when you subtract.

c. $f(n + 1) - f(n) = 3an^2 + (3a + 2b)n + (a + b + c)$

d. $a = \frac{1}{3}, b = \frac{1}{2}, c = \frac{1}{6}$; $0^2 + 1^2 + 2^2 + 3^2 + \ldots + n^2 = \frac{1}{3}n^3 + \frac{1}{2}n^2 + \frac{1}{6}n$

Quiz 1

1. $\frac{9}{4}$ **2.** $\frac{1}{16}$ **3.** $\frac{1}{16}$ **4.** 2 **5.** $\frac{1}{64x^9y^{15}}$

6. $\frac{x^{10}}{y^7}$ **7.** $\frac{y^6}{4x^4}$ **8.** x^9y^2

9.

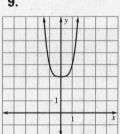

10.

11. $-13x^2 + 10x + 6$

12. $3x^3 + 11x^2 + 5x - 3$

13. $x^3 - 12x^2 + 48x - 64$ **14.** $9x^4 + 12x^2 + 4$

Lesson 6.4

Warm-ups

1. $4x(x - 6)$ **2.** $(2x - 3)(x + 7)$

3. $(2x - 9)^2$ **4.** -5 **5.** $\frac{3}{2}, -\frac{5}{3}$

Daily Homework Quiz

1. $4x^4 - 4x^3 + 15x^2 - 9$

2. $x^4 + 17x^3 - 11x^2 + 6$

3. $4x^2 - 16x + 15$

4. $-4x^4 + 5x^3 - x^2 - 10x + 10$

5. $125x^3 + 150x^2y + 60xy^2 + 8y^3$

6. $10x^3 - 61x^2 + 3x + 18$

Lesson Opener
Allow 20 minutes.

1. $x(x - 2)(x + 8) = 96$ **2.** *Sample answer:* The other side of the equation is not 0, so we cannot use the zero product property.

3. $x^3 + 6x^2 - 16x - 96 = 0$ **4.** $x^2(x + 6)$

5. $-16(x + 6)$ **6.** *Sample answer:* They both contain $x + 6$ as a factor.

7. $(x^2 - 16)(x + 6) = 0$ which factors further as $(x - 4)(x + 4)(x + 6) = 0$; $-6, -4, 4$ **8.** 4

9. 2 cm by 4 cm by 12 cm

Practice A

1. D **2.** C **3.** E **4.** F **5.** A **6.** B **7.** G

8. $(x + 1)(x^2 - x + 1)$

9. $(x + 3)(x^2 - 3x + 9)$

10. $(x + 5)(x^2 - 5x + 25)$

11. $(x - 1)(x^2 + x + 1)$

12. $(x - 2)(x^2 + 2x + 4)$

13. $(x - 4)(x^2 + 4x + 16)$ **14.** $(x + 3)(x^2 + 2)$

15. $(x - 1)(x^2 + 4)$ **16.** $(x + 5)(x^2 + 1)$

17. $(x - 6)(x^2 + 1)$ **18.** $(x + 4)(x^2 + 3)$

19. $(x - 5)(x^2 + 2)$ **20.** $-2, 0$ **21.** $0, 3$

22. $-4, 1$ **23.** $-3, -2$ **24.** $-7, 7$

25. $-10, 10$ **26.** $-2, -1, 1$ **27.** $-2, 1, 2$

28. $-3, -1, 3$ **29.** C **30.** A **31.** B

Practice B

1. $(x + 4)(x^2 - 4x + 16)$

2. $(x + 6)(x^2 - 6x + 36)$

3. $(x - 10)(x^2 + 10x + 100)$

4. $(x - 7)(x^2 + 7x + 49)$

5. $(2x - 1)(4x^2 + 2x + 1)$

6. $(2x - 5)(4x^2 + 10x + 25)$

7. $(3x - 2)(9x^2 + 6x + 4)$

8. $(2x + 10)(4x^2 - 20x + 100)$

9. $(3x + 8)(9x^2 - 24x + 64)$

10. $(4x + 3)(16x^2 - 12x + 9)$

11. $(10x - 1)(100x^2 + 10x + 1)$

12. $(5x + 4)(25x^2 - 20x + 16)$

13. $(x - 3)(x^2 + 5)$ **14.** $(x - 4)(x^2 + 2)$

15. $(x + 2)(x^2 + 7)$ **16.** $(x - 4)(3x^2 + 2)$

17. $(x + 1)(5x^2 + 1)$ **18.** $(x - 6)(2x^2 + 5)$

19. $(x - 2)^2(x + 2)$ **20.** $(x - 5)(x - 3)(x + 3)$

21. $(x + 1)(x - 4)(x + 4)$

22. $(x - 1)(2x - 3)(2x + 3)$

23. $(x - 3)(4x - 1)(4x + 1)$

24. $(x + 2)(3x - 2)(3x + 2)$

25. $(2x - 3)(2x + 3)(4x^2 + 9)$

26. $(x^2 - 3)(x^2 + 3)$ **27.** $(x^2 + 3)(x^2 + 2)$

28. $(x^2 + 3)(x^2 - 2)$ **29.** $(x^2 - 3)(x^2 + 8)$

30. $(x^2 - 5)(x^2 - 2)$ **31.** $2x^2(x - 10)(x + 10)$

32. $2x^2(2x - 3)(2x + 3)$

33. $3x^2(3x - 1)(3x + 1)$

34. $3(x - 1)(x + 1)(x^2 + 1)$

35. $2(x^2 + 2)(x^2 + 6)$ **36.** $-(x^2 + 7)(x^2 + 3)$

37. $-6, -2, 2$ **38.** 2 **39.** $-\frac{3}{2}$ **40.** $\frac{1}{3}$ **41.** -7

42. $-3, 3, 5$ **43.** $-2, 2$ **44.** No real solutions

45. $-3, -1, 1, 3$ **46.** $-\sqrt{5}, \sqrt{5}$

47. $-\sqrt{6}, -2, 2, \sqrt{6}$

48. $-2\sqrt{2}, -\sqrt{2}, \sqrt{2}, 2\sqrt{2}$ **49.** 750 ft³

50. $x^3 - 15x^2 + 50x = 750$ **51.** 15

52. 10 ft by 15 ft by 5 ft

Practice C

1. $(x + 9)(x^2 - 9x + 81)$

2. $(4x - 5)(16x^2 + 20x + 25)$

3. $2(x + 2)(x^2 - 2x + 4)$

4. $5(2x + 1)(4x^2 - 2x + 1)$

5. $2(2x - 3)(4x^2 + 6x + 9)$

6. $(2x - 3)(2x + 3)(4x^2 + 9)$

7. $(x^2 - 8)(x^2 + 8)$

8. $3(x - 2)(x + 2)(x^2 + 4)$

9. $(x + 2)(x - 7)(x + 7)$ **10.** $(x - 1)(x^2 + 3)$

11. $(x - 6)(2x - 1)(2x + 1)$

12. $(x + 2)(x - 1)(x^2 + x + 1)$

13. $(x + 1)(2x - 1)(4x^2 + 2x + 1)$

14. $(x + 5)(x + 1)(x^2 - x + 1)$

15. $8(x + 1)(x + 2)(x^2 - 2x + 4)$

16. $3(x + 4)(x - 1)(x + 1)$

Algebra 2
Chapter 6 Resource Book

17. $3(x + 2)(x - 2)(x^2 + 2x + 4)$

18. $x(x - 3)(x - 2)(x + 2)$

19. $x^3(x + 1)^2(x - 1)$ **20.** $6x(x + 2)(x^2 + 2)$

21. $x^2(x + 1)(x^4 + 1)$ **22.** 11   **23.** $-\frac{2}{3}$ **24.** 3

25. $-2, 2$ **26.** $-\frac{1}{2}, \frac{1}{2}$ **27.** $-2, 2, 3$ **28.** -8

29. $-3, -\frac{2}{3}, \frac{2}{3}$ **30.** 2, 5   **31.** $1, -\frac{2}{3}$ **32.** 3

33. $-1, 5$ **34.** $-6, -3, 0, 3$ **35.** $-1, 0, 1$

36. $-3, 0, 2$ **37.** $-2, 0, 1, 2$ **38.** $-1, 0, 1, 3$

39. $-\sqrt{3}, \sqrt{3}, 2$ **40.** $-4, -\sqrt{7}, \sqrt{7}$

41. $-\sqrt{\frac{5}{2}}, \sqrt{\frac{5}{2}}, 1$ **42.** $-\sqrt{6}, \sqrt{6}$

43. $-\sqrt{10}, 0, \sqrt{10}$ **44.** $-\sqrt{2}, \sqrt{2}$ **45.** $-2, 2$

46. 99 in.^{3}

Reteaching with Practice

1. $(x + 5)(x^2 - 5x + 25)$

2. $(x - 7)(x^2 + 7x + 49)$

3. $(4x - 1)(16x^2 + 4x + 1)$

4. $(2x + 3)(4x^2 - 6x + 9)$

5. $3(x - 2)(x^2 + 2x + 4)$

6. $(10x - 9)(100x^2 + 90x + 81)$

7. $(x + 3)(x - 3)(x - 1)$

8. $(x + 1)(x - 1)(x + 5)$

9. $(x + 4)(x - 4)(x - 3)$

10. $(5x^2 + 3)(5x^2 - 3)$ **11.** $(x^2 - 3)(x^2 + 2)$

12. $(x^2 - 8)^2$ **13.** $(7x^2 + 2)(7x^2 - 2)$

14. $0, 3$ **15.** $-3, 5$ **16.** $-2, 0, 2$ **17.** $-6, 4$

Cooperative Learning Activity

Instructions

1. -12 **2.** 4, 1   **3.** $-2, 1, -1$

4. $1, -1, -2, \frac{3}{2}$ **5.** $1, 2, -2, \sqrt{3}, -\sqrt{3}$

Analyzing the Results

1. Answers will vary   **2.** Yes   **3.** The number of solutions of a polynomial equation is equal to the degree of the equation.

Interdisciplinary Application

1. $100x^3 - 300x^2 + 176x - 528 = 0$ **2.** 3

3. height = 3 yards; width = 11 yards; length = 16 yards

4. height = 3 yards; width = 7 yards; length = 29 yards

Math and History Applications

1. $(x + 5)x - 3$

2. **a.** $([(x - 7)x + 1]x - 1)x + 4$

   b. 3 multiplications, 4 additions or subtractions

   c. 7   **3.** $[(4x)x - 2]x + 1$ **4.** $n - 1$

Challenge: Skills and Applications

1. $(x - 2)^2(x + 1)$ **2.** $x(x + 3)^3(x - 3)$

3. **a.** $(a^2 + 1)(a^4 - a^2 + 1)$

   b. $65 = 5 \cdot 13; 4097 = 17 \cdot 241$

   c. $(2^{2k} - 2^k + 1)$, and if $k > 0$, both factors are greater than 1, so this is a factorization of $2^n + 1$ which shows that $2^n + 1$ is not prime.

4. **a.** $(x^2 - 4)(x^4 + 4x^2 + 16) =$
   $(x - 2)(x + 2)(x^4 + 4x^2 + 16)$

   b. $(x^3 - 8)(x^3 + 8) = (x - 2)(x^2 + 2x + 4)$
   $(x + 2)(x^2 - 2x + 4)$

   c. $x^4 + 4x^2 + 16 = (x^2 + 2x + 4)(x^2 - 2x + 4)$

   d. $x^4 + a^2x^2 + a^4 = (x^2 + ax + a^2)$
   $(x^2 - ax + a^2)$

5. **a.** $(x^2 + 4i)(x^2 - 4i)$ **b.** $(x + pi)(x - pi)$;
   yes   **c.** $w^2 = \frac{1}{2}p^2(1 + i)^2 = \frac{1}{2}p^2(2i) = p^2i$

   d. $(x + (i - 1)\sqrt{2})(x - (i - 1)\sqrt{2})$
   $(x - 2\sqrt{i})(x + 2\sqrt{i})$

Lesson 6.5

Warm-ups

1. $3x$ **2.** 5   **3.** $x + 7$ **4.** $2x + 3$

Daily Homework Quiz

1. $7x$ **2.** $15x^2$ **3.** $(2x + 5)(4x^2 - 10x + 25)$

4. $(x + 2)(5x^2 - 1)$ **5.** $2x^4(10x + 1)(10x - 1)$

Lesson 6.5 *continued*

Lesson Opener

Allow 15 minutes.

1.

$$12 \overline{)3105} $$

Quotient: 258

$$\begin{array}{r} 258 \\ 12\overline{)3105} \\ 24 \\ \hline 70 \\ 60 \\ \hline 105 \\ 96 \\ \hline 9 \end{array}$$

2. a. 3105 **b.** 12 **c.** 258 **d.** 9

3. *Sample answer:* Check that $258 \cdot 12 + 9$ equals 3105. **4.** *Sample answer:* Multiply the divisor by a digit from the quotient.

5. *Sample answer:* $315 \div 12$ is 26, remainder 3. The zero is important because it helps to indicate that in 3105 the 3 represents 3 thousand and the 1 represents 1 hundred.

Practice A

1. Dividend: $x^3 - 2x^2 - 14x - 5$, Divisor: $x - 5$, Quotient: $x^2 + 3x + 1$, Remainder: 0

2. Dividend: $2x^3 + 3x^2 + 3x + 17$, Divisor: $x + 2$, Quotient: $2x^2 - x + 5$, Remainder: 7

3. Dividend: $x^3 + x - 2$, Divisor: $x - 3$, Quotient: $x^2 + 3x + 10$, Remainder: 28

4. $x + 2 - \dfrac{8}{x + 1}$ **5.** $x + 3 + \dfrac{3}{x - 2}$

6. $x + 2 - \dfrac{4}{x + 3}$ **7.** $x - 6 - \dfrac{5}{x - 1}$

8. $x + 2$ **9.** $x + 2 + \dfrac{5}{x - 5}$ **10.** $x + 2$

11. $x - 4 + \dfrac{5}{x + 1}$ **12.** $x - 3 - \dfrac{3}{x - 2}$

13. $x + 4 - \dfrac{8}{x + 3}$ **14.** $x + 1$

15. $x + 5 + \dfrac{8}{x - 2}$ **16.** $x + 6 - \dfrac{15}{x + 1}$

17. $x - 1 + \dfrac{6}{x - 2}$ **18.** $x - 2$

19. $x + 5 - \dfrac{2}{x + 1}$ **20.** $x - 6$

21. $x + 1 + \dfrac{3}{x - 4}$ **22.** $x + 7$ **23.** $x - 2$

24. $x + 3$

25.

$$\begin{array}{r|rrr} 3 & 1 & -6 & -1 \\ & & 3 & -9 \\ \hline & 1 & -3 & -10 \end{array} = x - 3 - \dfrac{10}{x - 3}.$$

The denominator of the remainder is $x - 3$, not $x + 3$.

26. As written, synthetic division cannot be used because the divisor does not have the form $x - k$.

Practice B

1. $x + 5 + \dfrac{21}{x - 3}$ **2.** $2x + 3$ **3.** $x^2 + x + 1$

4. $2x^2 + x - 3 + \dfrac{5}{2x - 1}$

5. $x^2 + \dfrac{1}{3}x - \dfrac{16}{9} + \dfrac{25}{9(3x + 1)}$

6. $4x + \dfrac{7}{2} + \dfrac{23}{2(2x - 3)}$

7. $x + 2 - \dfrac{1}{x^2 + 3x - 1}$

8. $x - 4 + \dfrac{2(6x - 11)}{x^2 + x - 4}$

9. $2x^2 + x + 3$ **10.** $x^2 - 3x + 7 - \dfrac{9}{x + 3}$

11. $x^3 + 4 + \dfrac{3}{x - 5}$

12. $3x^2 - 8x + 21 - \dfrac{43}{x + 2}$

13. $5x^3 + 3x^2 + 5 + \dfrac{6}{x - 1}$

14. $3x^3 - 10x^2 + 40x - 160 + \dfrac{635}{x + 4}$

15. $x^2 - x + 1 - \dfrac{3}{x + 1}$

16. $3x^3 + 6x^2 + 12x + 24 + \dfrac{47}{x - 2}$

17. $-8, 2$ **18.** 2 **19.** $-\dfrac{1}{2}, 1$ **20.** $-\dfrac{2}{3}, 1$

21. $x - 2$ **22.** $P(x) = 50x - 5x^3$

23. ≈ 0.3 million

Practice C

1. $x - 6 + \dfrac{3(7x - 4)}{x^2 + 3x - 1}$

Lesson 6.5 *continued*

2. $2x^2 - 4x + 9 - \dfrac{28}{2x + 3}$

3. $2x^2 - 4x + 6 - \dfrac{5x + 7}{x^2 + 2x + 1}$

4. $4x - 6 + \dfrac{16x - 19}{x^2 - 4}$

5. $2x - \dfrac{4}{3} + \dfrac{4x + 15}{3(3x^2 + x)}$

6. $\dfrac{1}{3}x^2 - \dfrac{8}{9}x + \dfrac{61}{27} - \dfrac{149}{27(3x + 2)}$

7. $\dfrac{1}{2}x + \dfrac{x + 10}{2(2x^2 - 1)}$

8. $\dfrac{1}{2}x - \dfrac{5}{4} + \dfrac{3(8x + 5)}{4(4x^2 + 2x - 1)}$

9. $5x^3 + 8x^2 + 23x + 52 + \dfrac{96}{x - 2}$

10. $6x^2 - 20x + 65 - \dfrac{192}{x + 3}$

11. $2x^2 + 2x - 1 + \dfrac{3}{x - 1}$

12. $4x^2 - 10x + 20 - \dfrac{39}{x + 2}$

13. $3x^4 + 9x^3 + 29x^2 + 87x + 256 + \dfrac{769}{x - 3}$

14. $2x^2 + 5x - 1$ **15.** $-1, 7$ **16.** $-\dfrac{3}{2}, \dfrac{1}{3}$

17. $-2 - \sqrt{5}, -2 + \sqrt{5}$

18. $-\dfrac{5 - \sqrt{17}}{2}, -\dfrac{5 + \sqrt{17}}{2}$ **19.** $-5, -1$

20. $\dfrac{3 - \sqrt{5}}{2}, \dfrac{3 + \sqrt{5}}{2}$ **21.** $-\dfrac{3}{2}, 1$

22. $\dfrac{1 - \sqrt{10}}{3}, \dfrac{1 + \sqrt{10}}{3}$ **23.** $-1 - i, -1 + i$

24. $-\sqrt{5}\,i, \sqrt{5}\,i$

25. $A = (-0.004t^3 + 0.082t^2 - 0.268t + 3.206) \div (2.61t + 247);\ 0.0118\ \dfrac{\text{quadrillion Btu}}{\text{million people}}$

26. 810 yearbooks

Reteaching with Practice

1. $2x - 13 + \dfrac{66}{x + 5}$ **2.** $3x + 2 + \dfrac{6}{x - 1}$

---

3. $x - 4 + \dfrac{14}{x + 2}$ **4.** $x + 3 + \dfrac{17}{x - 8}$

5. $(x - 3)$ **6.** $(2x + 1)$ **7.** $x - 2 + \dfrac{-7}{x + 4}$

8. $(3x + 4)$ **9.** $(x - 3)(x + 2)(x + 3)$

10. $(x + 4)(x - 2)(x - 1)$

Real-Life Application

1. Sales increase as price decreases and vice versa. **2.** $P = -2x^3 + 25x$

3.

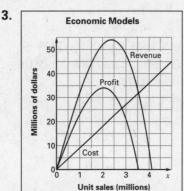

4. 1,550,000 units **5.** $21.50 per unit

6.

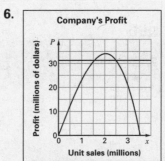

7. $26.096 million **8.** Yes, producing 2 million units sold at $26 per unit will give a profit of $34 million.

Challenge: Skills and Applications

1. 120 **2.** 16

3. a. $i^4 + i^3 - 5i^2 + i - 6 = 1 - i - 5(-1) + i - 6 = 0$;
$(-i)^4 + (-i)^3 - 5(-i)^2 + (-i) - 6 = 1 + i - 5(-1) - i - 6 = 0$ **b.** $-3, 2$; no

Lesson 6.5 *continued*

4. a. $P(b) = (b - b)Q(b) + R = 0 \cdot Q(b) + R = R$; this proves the Remainder Theorem. **b.** According to the result of part (a), $R = 0$ if and only if $P(b) = 0$. That is, the remainder of division by $x - b$ is 0 if and only if b is a zero of $P(x)$. But the remainder is 0 if and only if $x - b$ is a factor of $P(x)$.

5. a. $\dfrac{1}{1 - x} = 1 + x + x^2 + x^3 + x^4 + \ldots$

b. $\dfrac{1}{1 - \frac{1}{2}} = 2; \ 1 + \dfrac{1}{2} + \dfrac{1}{4} + \dfrac{1}{8} + \ldots$

≈ 1.875; as more terms are added, they get closer to being equal.

c. $\dfrac{1}{(1 - x)^2} = 1 + 2x + 3x^2 + 4x^3 + \ldots$

Lesson 6.6

Warm-ups

1. $(3x - 4)(2x + 7)$ **2.** $(5x^2 - 2x + 4)(x - 4)$
3. $-2, \frac{1}{2}$ **4.** $-2, 2, -\sqrt{5}, \sqrt{5}$ **5.** 0

Daily Homework Quiz

1. $x - 12 + \dfrac{48}{x + 5}$ **2.** $2x^2 + 2x - 4 - \dfrac{1}{x + 2}$

3. $(x - 5)(x + 4)(x - 1)$ **4.** 2, 7

Lesson Opener

Allow 15 minutes.

1. *Sample answer:*
$(3x - 5)(7x + 2)(3x - 2) =$
$(3x - 5)(21x^2 - 8x - 4) =$
$3x(21x^2 - 8x - 4) - 5(21x^2 - 8x - 4) =$
$(63x^3 - 24x^2 - 12x) - (105x^2 - 40x - 20) =$
$63x^3 - 129x^2 + 28x + 20$

2. $\dfrac{5}{3}, -\dfrac{2}{7}, \dfrac{2}{3}$ **3.** factors **4.** $\dfrac{5}{3}, -\dfrac{2}{7}, \dfrac{2}{3}$

5. factors **6.** $-\dfrac{5}{2}, \dfrac{7}{3}$ **7.** denominators, factors, 6; numerators, factors, -35

Practice A

1. ± 1 **2.** $\pm 1, \pm 7$ **3.** $\pm 1, \pm 2, \pm 3, \pm 6$
4. $\pm 1, \pm 3, \pm 9$ **5.** $\pm 1, \pm 2, \pm 3, \pm 4, \pm 6, \pm 12$

6. $\pm 1, \pm 2, \pm 4, \pm 5, \pm 10, \pm 20$
7. $\pm 1, \pm 2, \pm 3, \pm 4, \pm 6, \pm 8, \pm 12, \pm 24$
8. $\pm 1, \pm 2, \pm 4, \pm 5, \pm 8, \pm 10, \pm 20, \pm 40$
9. $\pm 1, \pm 2, \pm 4, \pm 5, \pm 10, \pm 20, \pm 25, \pm 50, \pm 100$
10. -1 **11.** 1 **12.** neither **13.** 1 **14.** neither
15. -1 and 1 **16.** -1 and 1 **17.** neither **18.** 1
19. $-3, -2, 4$ **20.** $-1, 1, 2$ **21.** $-3, -2, 2$
22. $-1 - \sqrt{6}, -1, -1 + \sqrt{6}$
23. $-\sqrt{5}, \sqrt{5}, 3$ **24.** $x^3 + 5x^2 + 4x = 84$
25. $\pm 1, \pm 2, \pm 3, \pm 4, \pm 6, \pm 7, \pm 12, \pm 14, \pm 21, \pm 28,$
$\pm 42, \pm 84$ **26.** 3 in. by 4 in. by 7 in.

Practice B

1. $\pm 1, \pm 2, \pm 4$ **2.** $\pm 1, \pm 2, \pm 3, \pm 6, \pm \frac{1}{2}, \pm \frac{3}{2}$
3. $\pm 1, \pm 2, \pm 4, \pm 8, \pm \frac{1}{3}, \pm \frac{2}{3}, \pm \frac{4}{3}, \pm \frac{8}{3}$
4. $-2, \frac{1}{2}, 3$ **5.** $-\frac{8}{3}, -1, 1$ **6.** $-\frac{3}{2}, \frac{1}{4}, 2$
7. $-2, 1, 5$ **8.** $-3, -1, -\sqrt{2}, \sqrt{2}$
9. $-3, -1, \frac{1}{2}, 1$ **10.** $-\sqrt{2}, \sqrt{2}, \frac{5}{2}$ **11.** $-\frac{3}{2}, \frac{1}{2}, 3$
12. $-\sqrt{3}, \sqrt{3}, 3$ **13.** $-2, -\sqrt{2}, \sqrt{2}, \frac{1}{2}$
14. $-3, -\sqrt{2}, 1, \sqrt{2}$
15. $-3, 1, -\dfrac{1 + \sqrt{17}}{4}, -\dfrac{1 - \sqrt{17}}{4}$
16. $t^3 - 13t^2 + 65t - 105 = 0$
17. $\pm 1, \pm 3, \pm 5, \pm 7, \pm 15, \pm 21, \pm 35, \pm 105$
18. 1, 3, 5, 7 **19.** 1983

Practice C

1. $-\frac{3}{2}, -\frac{1}{2}, 4$ **2.** $-4, -1, \frac{2}{3}, 1$ **3.** $-1, -\frac{1}{5}, \frac{5}{2}, 3$
4. $-\frac{3}{4}, -\frac{1}{2}, \frac{1}{3}, 2$ **5.** $-3, -\frac{7}{2}, -\frac{1}{3}, 1$
6. $-\frac{5}{2}, -\frac{3}{2}, \frac{1}{2}$ **7.** $-7, \dfrac{5 - \sqrt{21}}{2}, \dfrac{5 + \sqrt{21}}{2}$
8. $4 - \sqrt{14}, \frac{1}{6}, 4 + \sqrt{14}$
9. $-1 - \sqrt{6}, -\frac{4}{3}, -1 + \sqrt{6}$
10. $-3 - 2\sqrt{2}, -1, -3 + 2\sqrt{2}, 3$
11. $-\dfrac{5 - \sqrt{17}}{4}, -1, -\frac{2}{3}, -\dfrac{5 + \sqrt{17}}{4}$
12. $-\dfrac{5}{2}, -\dfrac{1 - \sqrt{13}}{6}, -\dfrac{1 + \sqrt{13}}{6}, \dfrac{1}{2}$
13. $-\frac{9}{2}, -3, -\frac{5}{3}, 4$ **14.** $-\frac{7}{2}, -3, -\frac{5}{2}, \frac{1}{2}$
15. a_0 cannot be 0. **16.** $x(x^3 - x^2 - 24x - 36)$
17. $-3, -2, 6$ **18.** $-2, -\frac{1}{3}, 0, 2$
19. $-2, -1, 1$ **20.** The zeros of $f(x)$ are also the

zeros of $af(x)$.   **21.** To apply the rational zero theorem, the coefficients must be integers.

22. $2f(x) = x^3 - 19x - 30; -3, -2, 5$

Reteaching with Practice

1. $-3, 1, 2$ **2.** $-1, -2, 2$ **3.** $-4, 1, 2$
4. $-1, -\frac{1}{2}, 3$ **5.** $-3, \pm\sqrt{5}$ **6.** $2, -\frac{1}{2}, \pm\sqrt{6}$
7. $1, 2 \pm \sqrt{3}$ **8.** $-1, 1 \pm \sqrt{2}$

Interdisciplinary Application

1. 18,750 gallons

2. Aquarium 1:
$x^3 - 96x^2 + 2304x - 41,472 = 0$; length $= 72$ inches, width $= 24$ inches, height $= 24$ inches

Aquarium 2: $x^3 - 11.5x^2 + 33x - 40 = 0$;
length $= 8$ feet, width $= 2$ feet, height $= 2.5$ feet

Aquarium 3: $3x^3 + 12x^2 - 28,800 = 0$;
length $= 60$ inches, width $= 24$ inches,
height $= 20$ inches

Challenge: Skills and Applications

1. 1   **2.** −1   **3. a.** $a = c, b = d$ **b.** yes
4. a. $0, \pm 3, \pm 6, \pm 9$ **b.** $6; -1, \sqrt{2}, -\sqrt{2}$
5. a. $\pm 1, \pm 2$ **b.** By substitution, it is easily shown that none of these four numbers is a zero of the equation. Therefore, by the rational zero theorem, it can have no rational zeros.

6. a. $ap^3 = -bp^2q - cpq^2 - dq^3$

b. The right-hand side can be written as $q(-bp^2 - cpq - dq^2)$, and so must be divisible by q. Therefore, the left-hand side is also divisible by q. But, because q is prime, it must divide a or p^3. Since p is not divisible by q, q must divide a.

c. $dq^3 = -ap^3 - bp^2q - cpq^2$; this shows that the right-hand side must be divisible by p, hence so is the left-hand side. But since p is prime it must divide d or q^3. Since p does not divide q^3, it must divide d.

Quiz 2

1. $3(x + 3)(x^2 - 3x + 9)$
2. $4(x^2 + 3)(x + 3)$ **3.** $5(x^2 + 3)(x^2 - 3)$
4. $\pm\sqrt{5}, \pm 2$ **5.** $x^2 + 2x + 13 + \dfrac{59}{x - 5}$

6. $2x - 3 + \dfrac{-4}{4x + 1}$

7. $x + 6 + \dfrac{-10x^2 - 3x - 31}{x^3 + 5}$ **8.** $-2, 1, \dfrac{3}{2}$

Lesson 6.7

Warm-ups

1. 3   **2.** $-\frac{5}{2}, 3$ **3.** 1   **4.** $\frac{1}{2}$ **5.** none

Daily Homework Quiz

1. $\pm 1, \pm 2, \pm 4, \pm\frac{1}{2}, \pm\frac{1}{3}, \pm\frac{2}{3}, \pm\frac{1}{6}, \pm\frac{4}{3}$
2. $-3, -2, 3$ **3.** $-\sqrt{5}, -\frac{3}{2}, \sqrt{5}$ **4.** $\frac{1}{2}, \frac{2}{3}, 5$

Lesson Opener

Allow 20 minutes.

1. 1 solution   **2.** 1 solution   **3.** no solution
4. 2 solutions   **5.** 2 solutions   **6.** 1 solution
7. 3 solutions   **8.** 3 solutions   **9.** 2 solutions
10. 4 solutions   **11.** 1 solution
12. 0 or 2 solutions   **13.** 1 or 3 solutions
14. 0, 2, or 4 solutions
15. The possibilities are $0, 2, 4, \ldots, n$.
16. The possibilities are $1, 3, 5, \ldots, n$.

Practice A

1. 3   **2.** 5   **3.** 4   **4.** 6   **5.** 3   **6.** 2   **7.** 1
8. 5   **9.** $2 - i$ **10.** $3 + 5i$ **11.** $-6 + 2i$
12. $-7 - 3i$ **13.** $\sqrt{2} - i$ **14.** $\sqrt{5} + 4i$
15. $3 - \sqrt{2}i$ **16.** $-2 - \sqrt{3}i$ **17.** $\sqrt{3} + \sqrt{5}i$
18. yes   **19.** no   **20.** no   **21.** yes   **22.** yes
23. no   **24.** $f(x) = x - 6$
25. $f(x) = x^2 + x - 6$ **26.** $f(x) = x^2 + 6x + 5$
27. $f(x) = x^3 - 3x^2 - x + 3$
28. $f(x) = x^3 - 7x^2 + 12x$
29. $f(x) = x^3 - 8x^2 + x + 42$
30. $(x + 3)(x + 1)(x - 2)$
31. Length: $(x + 3)$; Width: $(x + 1)$;
Height: $(x - 2)$ **32.** 10

Practice B

1. 3   **2.** 6   **3.** 5   **4.** 4   **5.** yes   **6.** no   **7.** yes
8. yes   **9.** $(x - 3), (x - 1), (x - 2)$
10. $(x - 4), (x + 1), (x + 2), x$

11. $(x + 6), (x - 2), (x - 1), (x - 1)$

12. $(x + 3), (x - i), (x + i)$

13. $(x - 4), (x + 5), (x - 2i), (x + 2i)$

14. $(x - 3), (x - (2 + i)), (x - (2 - i))$

15. $f(x) = x^3 - 5x^2 + 2x + 8$

16. $f(x) = x^3 - 5x^2 + 7x - 3$

17. $f(x) = x^3 + x^2 - 6x$

18. $f(x) = x^3 + 2x^2 + x + 2$

19. $f(x) = x^4 - x^3 + 9x^2 - 9x$

20. $f(x) = x^4 - 3x^3 + 3x^2 - 3x + 2$

21. $f(x) = x^4 + 10x^2 + 9$

22. $f(x) = x^3 - 10x^2 + 33x - 34$

23. $f(x) = x^4 + 2x^2 + 8x + 5$ **24.** $-1, i, -i$

25. $-3, -\frac{1}{2}, 2$ **26.** $-4, 3i, -3i$

27. $-1, \frac{1}{2}, i, -i$ **28.** $-2, 2, i, -i$

29. $-4, \frac{1}{3}, 2i, -2i$ **30.** 1996 **31.** 5

Practice C

1. 4 **2.** 3 **3.** 5 **4.** no **5.** no **6.** no

7. no **8.** $f(x) = 2x^4 - 4x^3 - 32x^2 + 64x$

9. $f(x) = 2x^4 + 58x^2 + 200$

10. $f(x) = 2x^3 - 12x^2 + 2x + 68$

11. $f(x) = 2x^3 - 12x^2 + 50x$

12. $f(x) = 2x^4 - 14x^3 + 38x^2 - 46x + 20$

13. $f(x) = 2x^6 - 26x^5 + 120x^4 - 200x^3 + 118x^2 - 174x$ **14.** $7, 5 + i, 5 - i$

15. $1, -1, 2 + \sqrt{3}, 2 - \sqrt{3}$ **16.** -3

17. $2 + i\sqrt{3}, 2 - i\sqrt{3}, -2, 0$

18. $-i, i, -1 + \sqrt{2}, -1 - \sqrt{2}$

19. $1 + 4i, 1 - 4i, \sqrt{3}, -\sqrt{3}$

20.
$R = -315.035t^5 + 5562.592t^4 + 1832.426t^3 - 708,818.278t^2 + 6,449,569.245t + 49,245,170.73; 1993$ **21.** The fifth solution must be a repeated solution because complex solutions come in conjugate pairs.

Reteaching with Practice

1. 4 zeros **2.** 3 zeros **3.** 3 zeros **4.** 4 zeros

5. $2, -1, \pm i$ **6.** $1, 3, 4$ **7.** $2, \pm i\sqrt{5}$

8. $0, \pm 2, -3$ **9.** $f(x) = x^3 - 6x^2 + 3x + 10$

10. $f(x) = x^3 - 7x^2 - 6x + 72$

---

11. $f(x) = x^3 - 7x^2 - x + 7$

12. $f(x) = x^3 + 2x^2 - 9x - 18$

13. $-2.60, 0.34, 2.26$ **14.** 1 **15.** -0.52

16. $-1.53, -0.35, 1.88$

Real-Life Application

1.

t	4	5	6	7	8	9
S	5049	5119	5167	5200	5224	5247

t	10	11	12	13	14	15
S	5275	5316	5376	5463	5584	5746

t	16	17
S	5955	6218

1997

2.

High School Sports Participation

3. $1.169t^3 - 28.77t^2 + 258.1t + 4402 = 6000$;
$1.169t^3 - 28.77t^2 + 258.1t - 1598 = 0; t \approx 16$

4. $1.169t^3 - 28.77t^2 + 258.1t + 4402 = 7100$;
$1.169t^3 - 28.77t^2 + 258.1t - 2698 = 0; t \approx 19$

5. 2002; 2004 **6.** about 891,110 students

Challenge: Skills and Applications

1. a. $(x - r_1)(x - r_2)(x - r_3)$;
$b = -(r_1 + r_2 + r_3)$ **b.** The coefficient of x^{n-1} is the negative of the sum of the zeros.

2. a. If the graph is moved slightly downward, the function will intersect the x-axis 3 times; this will happen if the function is changed to $f(x) - k$ for a small positive k.

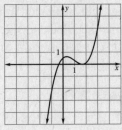

b. For a some positive value of m, the graph of $f(x) + m$ will intersect the x-axis only once. **3. a.** The graph intersects every horizontal line between the lines $y = A$ and $y = B$.

Lesson 6.7 *continued*

b. Let $C = 0$. Since $f(0) < 0 < f(2)$, there will be some t such that $f(t) = 0$. **c.** If $f(1) > 0$, $f(x)$ has a zero between $x = 0$ and $x = 1$; if $f(x) < 0$, $f(x)$ has a zero between 1 and 2.
d. Use $r = 0$, $s = 1$ in the first case; use $r = 1$, $s = 2$ in the second case. In either case, $f(x) > 0$ at the left endpoint of the new interval and $f(x) < 0$ at the right endpoint. Test the midpoint of the new interval and repeat the procedure, thus producing a sequence of intervals, each containing a zero of $f(x) = 0$ and each half as long as the preceding interval. **4. a.** $\overline{f(z)} = $ the conjugate of $az^n + bz^{n-1} + \cdots + jz + k = \overline{a} \cdot \overline{z}^n + \overline{b} \cdot \overline{z}^{n-1} + \cdots + \overline{j} \cdot \overline{z} + \overline{k}$, according to the facts previously proved. But the conjugates of the coefficients are the same as the coefficients themselves, since they are real. Therefore, $\overline{f(z)} = f(\overline{z})$. **b.** If you take the conjugate of both sides of the equation, you get $f(\overline{z}) = \overline{f(z)} = \overline{0} = 0$. This says that $\overline{z}$ is a zero of $f(x) = 0$.

Lesson 6.8

Warm-ups

1. $x - 2$ **2.** 3 **3.** minimum
4. $f(x) \to -\infty$ as $x \to -\infty$ and $f(x) \to +\infty$ as $x \to +\infty$.

Daily Homework Quiz

1. yes **2.** $-4, 2, \pm i$
3. $f(x) = x^3 + x^2 - 2x - 2$
4. $-0.73, 0.59, 2.73, 3.41$

Lesson Opener

Allow 10 minutes. Approximated answers are rounded to the nearest tenth.

1. $-1, 3; (1, -4)$ **2.** $-4, 2; (-1, -3)$
3. $-2, 0, 2; (-1.2, -3.1), (1.2, -3.1)$
4. $-3, -2, 1; (-2.5, 0.9), (-0.1, -6.1)$
5. $-2.8, 0, 2.8; (-2, -4), (0, 0), (2, -4)$
6. $0, 2; (1.3, -1.9)$

Graphing Calculator Activity

1. a.–e. For a general 4th degree function, the maximum number of distinct real zeros is 4, and the maximum number of turning points is 3.

Practice A

1. 4 **2.** 5 **3.** 3 **4.** $(-2, 5)$ local maximum; $\left(\frac{1}{2}, 3\right)$ local minimum **5.** $(1, 2)$ local minimum
6. $(-1, -4)$ local minimum; $\left(\frac{1}{2}, 1\right)$ local maximum; $(1, 1)$ local minimum

7.
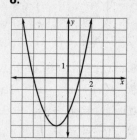
8.

9.
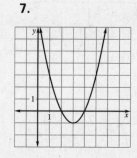
10.

11.
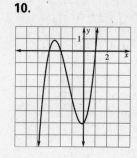
12.

13.
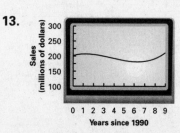

14. 1996 **15.** \$244 million

Practice B

1. 3 **2.** 2 **3.** 5 **4.** 1 **5.** 6 **6.** 8 **7.** B
8. C **9.** A **10.** $-3, 2, 5$ **11.** $-4, 6, 8$
12. $-3, 2$ **13.** $-5, -1, 7$ **14.** $-6, -2$ **15.** 8

Lesson 6.8 *continued*

16.

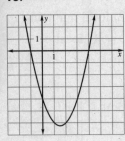

17.

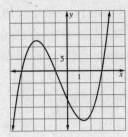

18.

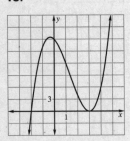

19.

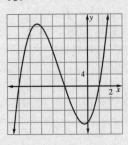

20.

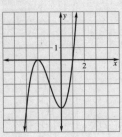

21.

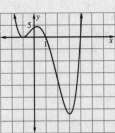

22.

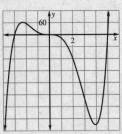

23.

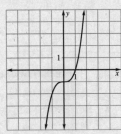

24.

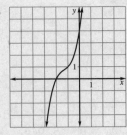

25. (8.78, 745.80) is a local maximum; (19.95, 652.46) is a local minimum; During the 1980 games the gold medal winner scored more points than in previous years and that was the record for several years following 1980. Starting in 1992, the number of points started to increase after several years of declining scores.

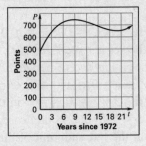

26. (8.5, 122,069.35) is a local maximum. (29.3, 96,068.15) is a local minimum. In 1973, the number of cattle on farms reached a maximum of 122,069.35 thousand. This number decreased to 96,068.15 in 1994 and then started to rise again.

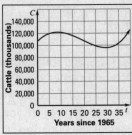

Practice C

1. 3 **2.** 2 **3.** 4

4.

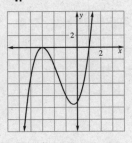

5.

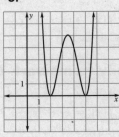

6.

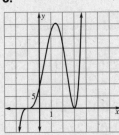

7.

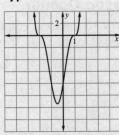

Lesson 6.8 *continued*

8. 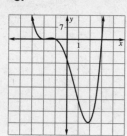   **9.**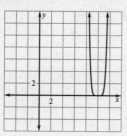

3.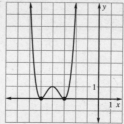

10. -1 **11.** 2, 4   **12.** 1, 3, 5

13. If n is even there is a turning point. If n is odd the graph passes through the x-axis.

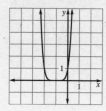   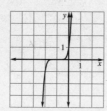

14. $-2, 3; 3$ **15.** $1, 7$; none   **16.** $-3, 5; -3, 5$

17. $-4, 3; -4$ **18.** $1, 3; 1$ **19.** $-4, 0, 2; -4$

20. $-3, 2, 5$; none   **21.** $-3, 2, 3; -3, 3$

22. $-8, 1, 4; 1$

23. 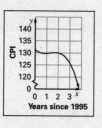 local minimum (0.86, 129.88), local maximum (1.72, 130.20); In 1996 you could buy more women's and girls' apparel with your money than in previous years, but by 1997 prices were higher than in recent years.

Reteaching with Practice

1. 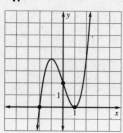   **2.**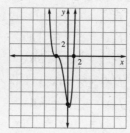

4. x-intercepts: $-2.85, 1.68$; local maximums: $(-2.05, 10.0)$, $(0.941, 7.03)$; local minimum: $(-0.388, 3.38)$ **5.** x-intercept: 2.32; local maximum: $(0.423, -0.615)$ local minimum: $(1.58, -1.38)$ **6.** x-intercepts: $-0.519, 1.29$; local minimum: $(0.606, -1.33)$ **7.** x-intercept: -1.65; local maximum: $(-0.707, 5.41)$; local minimum: $(0.707, 2.59)$ **8.** 1.57 inches; 67.6 cubic inches   **9.** About 4.86 inches by 8.86 inches by 1.57 inches

Interdisciplinary Application

1.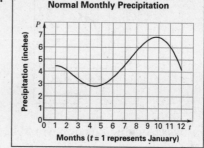

2. greatest amount occurs in September; least amount occurs in April   **3.** 54.95 inches

Challenge: Skills and Applications

1. The graph is translated right by 1 unit, left by 2 units, right by 4 units. In general the graph of $y = f(x - a)$ is a translation of the graph of $y = f(x)$, $|a|$ units to the right if a is positive, $|a|$ units to the left if a is negative.

2. a. At a turning point, the tangent line to the graph will be horizontal (i.e. it will have slope 0). Therefore, every turning point is a zero of $f'(x) = 0$.   **b.** At the point $(0, 0)$ on the graph of $y = x^3$ the tangent will have slope 0, but $(0, 0)$ is not a turning point.   **3. a.** $f'(x) = x^2 - 2x - 3$; turning points: $(3, -5)$ and $\left(-1, \frac{17}{3}\right)$

b. $f'(x) = x^3 - 4x$; turning points: $(0, 4)$, $(2, 0)$, $(-2, 0)$ **4. a.** $1; -1, 1$ **b.** 1

Lesson 6.8 *continued*

Lesson 6.9

Warm-ups

1. $f(x) = (x + 2)(x - 1)(x - 3)$ **2.** 17

3. $a = 3, b = -13, c = 19$

Daily Homework Quiz

1. 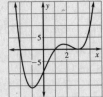 **2.**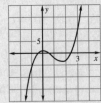

x-intercepts: about -0.41, 0.5, about 2.41; local max. at $(0, 1)$, local min. at about $(1.67, -3.63)$

Lesson Opener

Allow 10 minutes.

1, 3. *Sample answer:*

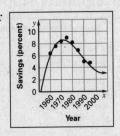

2. *Sample answer:* A cubic function would be best because the pattern appears to go up, then down, then about to go up again.

Practice A

1. B **2.** A **3.** C

4. $f(x) = (x + 1)(x - 1)(x - 3)$

5. $f(x) = (x + 3)(x + 2)(x - 1)$

6. $f(x) = (x + 3)(x + 1)(x - 4)$

7. $f(x) = (x + 3)(x + 1)(x - 6)$

8. $f(x) = (x + 2)(x - 3)(x - 5)$

9. $f(x) = (x - 3)(x - 4)(x - 5)$

10. $f(x) = (x + 2)(x + 1)(x - 6)$

11.

$f(1)$	$f(2)$	$f(3)$	$f(4)$	$f(5)$	$f(6)$	$f(7)$
0	0	2	6	12	20	30

0　　2　　4　　6　　8　　10

2　　2　　2　　2　　2

12.

$f(1)$	$f(2)$	$f(3)$	$f(4)$	$f(5)$	$f(6)$	$f(7)$
2	11	34	77	146	247	386

9　　23　　43　　69　　101　　139

14　　20　　26　　32　　38

6　　6　　6　　6

13.

$f(1)$	$f(2)$	$f(3)$	$f(4)$	$f(5)$	$f(6)$	$f(7)$
8	28	74	158	292	488	758

20　　46　　84　　134　　196　　270

26　　38　　50　　62　　74

12　　12　　12　　12

14. $y = 0.33x^3 - 3.25x^2 + 8.77x + 10.64$

15. $y = -0.22x^3 + 2.51x^2 - 8.98x + 20.43$

16. $y = -0.58x^3 + 5.07x^2 - 19.20x + 53.43$

17. $y = 0.28x^3 - 2.10x^2 + 5.56x + 1.79$

Practice B

1. $f(x) = (x + 2)(x + 1)(x - 2)$

2. $f(x) = 2(x + 1)(x - 1)(x - 3)$

3. $f(x) = \frac{1}{2}(x - 2)(x - 3)(x - 4)$

4. $f(x) = -2(x + 1)(x + 3)(x - 2)$

5. $f(x) = \frac{1}{2}(x + 4)(x + 1)(x - 5)$

6. $f(x) = -(x + 2)(x - 4)(x - 6)$

7. $f(x) = 2(x + 1)(x - 3)(x - 4)$

8. $f(x) = -2x(x - 1)(x - 8)$

9. $f(x) = \frac{1}{4}x(x - 3)(x - 9)$

Lesson 6.9 *continued*

10.

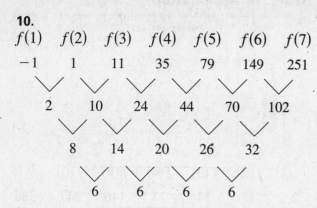

$f(1)$	$f(2)$	$f(3)$	$f(4)$	$f(5)$	$f(6)$	$f(7)$
-1	1	11	35	79	149	251

2   10   24   44   70   102

8   14   20   26   32

6   6   6   6

11.

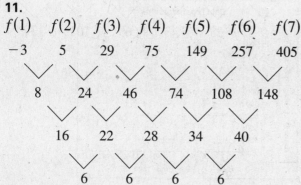

$f(1)$	$f(2)$	$f(3)$	$f(4)$	$f(5)$	$f(6)$	$f(7)$
-3	5	29	75	149	257	405

8   24   46   74   108   148

16   22   28   34   40

6   6   6   6

12.

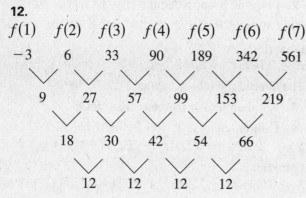

$f(1)$	$f(2)$	$f(3)$	$f(4)$	$f(5)$	$f(6)$	$f(7)$
-3	6	33	90	189	342	561

9   27   57   99   153   219

18   30   42   54   66

12   12   12   12

13. $f(x) = x^3 + 2x^2 + x + 1$

14. $f(x) = x^3 - 3x^2 + x - 4$

15. $f(x) = -x^2 + 3x + 2$

16. $f(x) = -x^3 + x^2 - 3x + 2$

17. $M = 0.000127t^3 - 0.00330t^2 + 0.0158t + 9.77$; 13.3 thousand miles

Practice C

1. $f(x) = (x + 5)(x + 2)(x - 1)$

2. $f(x) = -2(x + 2)(x - 1)(x - 2)$

3. $f(x) = (x + 1)(x - 2)^2$

4. $f(x) = \frac{3}{2}(x + 1)(x - 2)(x - 3)$

5. $f(x) = \frac{3}{2}(x + \frac{1}{2})(x - 1)(x - 3)$

6. $f(x) = -12(x - \frac{1}{2})(x - \frac{3}{2})(x - 2)$

7. $f(x) = 12(x + \frac{1}{3})(x - \frac{1}{4})(x - 1)$

8. $f(x) = (x + \frac{2}{3})(x + \frac{1}{4})(x - \frac{1}{2})$

9. $f(x) = 108(x + \frac{1}{3})(x - \frac{1}{3})(x - \frac{5}{6})$

10.

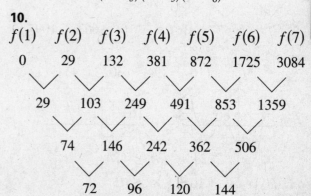

$f(1)$	$f(2)$	$f(3)$	$f(4)$	$f(5)$	$f(6)$	$f(7)$
0	29	132	381	872	1725	3084

29   103   249   491   853   1359

74   146   242   362   506

72   96   120   144

24   24   24

11.

$f(1)$	$f(2)$	$f(3)$	$f(4)$	$f(5)$	$f(6)$	$f(7)$
-2	9	76	265	666	1393	2584

11   67   189   401   727   1191

56   122   212   326   464

66   90   114   138

24   24   24

12.

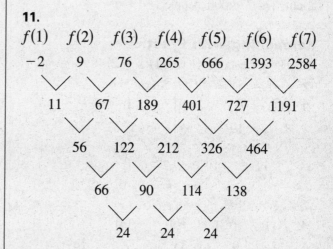

$f(1)$	$f(2)$	$f(3)$	$f(4)$	$f(5)$	$f(6)$	$f(7)$
3	22	213	966	3031	7638	16617

19   191   753   2065   4607   8979

172   562   1312   2542   4372

390   750   1230   1830

360   480   600

120   120

Lesson 6.9 *continued*

13.

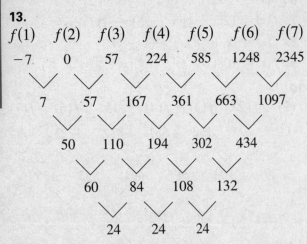

$f(1)$ $f(2)$ $f(3)$ $f(4)$ $f(5)$ $f(6)$ $f(7)$
-7 0 57 224 585 1248 2345

7 57 167 361 663 1097

50 110 194 302 434

60 84 108 132

24 24 24

14. $f(x) = x^3 - 10x^2 + 8x - 15$

15. $f(x) = x^3 + 8x^2 - 12x + 13$

16. $y = 1.114t^3 - 45.50t^2 + 2829.5t + 249,915$;
about $274,774,000$ people

Reteaching with Practice

1. $f(x) = -(x + 4)(x - 1)(x + 2)$

2. $f(x) = \frac{1}{2}(x + 2)(x - 5)(x + 1)$

3.

$f(1)$ $f(2)$ $f(3)$ $f(4)$
8 19 38 65

11 19 27

8 8

4.

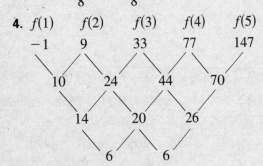

$f(1)$ $f(2)$ $f(3)$ $f(4)$ $f(5)$
-1 9 33 77 147

10 24 44 70

14 20 26

6 6

5. $f(x) = -x^2 + 5x + 4$

6. $f(x) = x^3 - 2x^2 - 11$

Real-Life Application

1.

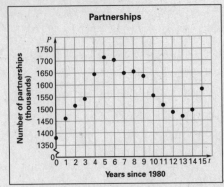

Partnerships

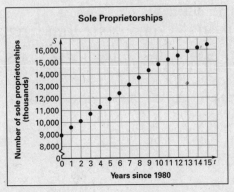

Sole Proprietorships

2. yes, sole proprietorships; $S = 517.7x + 9234$

3. Cubic: $P = 0.692t^3 - 19.20t^2 + 145.6t + 1335$;

Quartic:
$P = 0.1027t^4 - 2.388t^3 + 9.92t^2 + 55.4t + 1383$;
The quartic model **4.** about 11,773; 30,684

5. No, the number increases too rapidly.

6. Cubic:
$S = -1.348t^3 + 17.61t^2 + 532.3t + 8973$

Quartic:
$S = 0.0304t^4 - 2.261t^3 + 26.24t^2 + 505.6t + 8987$

7. No, both models decrease rapidly and eventually become negative. **8.** The number of sole proprietorships have been steadily increasing while the number of partnerships has fluctuated.

Challenge: Skills and Applications

1. a. 6, 0, 0, 6, 18, 36 **b.** The 3rd differences for *f* must be constant, but these are just the 2nd differences of the 1st differences. Therefore, the second differences of the values found in part (a) are constant, and by the second property of finite differences, these must be the values of a quadratic function; $3x^2 - 3x$ **c.** $-6, 0, 6, 12, 18$; $y = 6x$

Lesson 6.9 *continued*

2. a. $10, -6, 14$; $2a$ in all three cases

b. $-12, 30$; $6a$ in both cases **c.** $a(n!)$

3. a. The third-order difference depends only on the leading coefficient.

b. $f(x) = 3(x - 1)(x - 3)(x - 4)$ **c.** 6

4. a. 0 **b.** For any pair of consecutive terms c and $c + 1$, the first order difference is
$f(c) - f(c + 1) = [g(c) + h(c)] -$
$[g(c + 1) + h(c + 1)] = [g(c) - g(c + 1)] +$
$[h(c) - h(c + 1)]$. Similar arguments can be made for higher order differences. **c.** Suppose
$f(x) = ax^n + bx^{n-1} + \ldots + rx + s$. Let $g(x) = ax^n$ and $h(x) = bx^{n-1} + \ldots + rx + 5$. Then
$f(x) = g(x) + h(x)$ and the nth difference for $f(x)$ is equal to the nth difference of $g(x)$ added to the nth difference of $h(x)$. Since n is greater than the degree of $h(x)$, then nth difference for $h(x)$ is 0. Therefore the constant nth-order difference depends only on the first term. **d.** The given polynomial can be replaced by the polynomial ax^n.

Review and Assessment

Review Games and Activities

1. $\dfrac{125y}{3x^{22}}$ **2.** $f(x) = 0$

3. $x^3 + 2x^2 - 16x + 16$ **4.** $x = 5$

5. $x^3 + 4x^2 + 6x + 5 + \dfrac{7}{x - 1}$

6. $x = 3, 5, -2$ **7.** $x = \frac{1}{2}, -\frac{1}{2}, -2$

8. $(x - 3)(x^2 - 2x - 5)$

O I N K - M E N T
1 2 3 4 5 6 7 8

Test A

1. $\dfrac{1}{x}$ **2.** $8x^3y^3$ **3.** y^6 **4.** $-5x^2y$

5. $f(x) \to -\infty$ as $x \to -\infty$
and $f(x) \to +\infty$ as $x \to +\infty$;

x	-2	-1	0	1	2
y	-5	7	1	-5	7

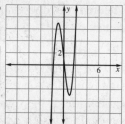

6. $f(x) \to +\infty$ as $x \to -\infty$
and $f(x) \to -\infty$ as $x \to +\infty$;

x	-2	-1	0	1	2
y	0	-3	0	3	0

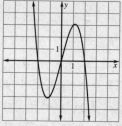

7. $2x^2 - 2x + 2$ **8.** $4x^2 - y^2$ **9.** $x^3 + 1$

10. $(5x + 1)(5x - 1)$ **11.** $(x + 1)(x^2 - x + 1)$

12. $4x^2y(3x^2y^2 + 5y - 6)$ **13.** $4, -4$

14. $2, -2, 3, -3$ **15.** $1, -1, -4$

16. $x^2 + 2x - 3$ **17.** $2x^2 + 8x + 8$

18. Possible zeros: $1, -1, 3, -3$; Zeros: $-1, -3$

19. Possible zeros: $1, -1, 2, -2, 4, -4, 8, -8$

Zeros: $-4, 1, 2$

20. $f(x) = x^3 + 2x^2 - 11x - 12$

21. $f(x) = x^2 - 7x + 12$ **22.** $-1.29, 2, 3.24$

23. x-intercepts: $(-3, 0), (3, 0)$;

local max: $(0, 27)$; local mins: $(-3, 0), (3, 0)$

The graph rises to the right and to the left.

24.

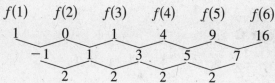

$f(1)$	$f(2)$	$f(3)$	$f(4)$	$f(5)$	$f(6)$
1	0	1	4	9	16

Test B

1. $\dfrac{y}{x}$ **2.** $\dfrac{1}{x^6y^9}$ **3.** x^8y^8 **4.** 1

5. $f(x) \to +\infty$ as $x \to -\infty$
and $f(x) \to -\infty$ as $x \to +\infty$;

x	-2	-1	0	1	2
y	8	1	0	-1	-8

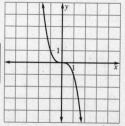

6. $f(x) \to -\infty$ as $x \to -\infty$
and $f(x) \to +\infty$ as $x \to +\infty$;

x	-2	-1	0	1	2
y	0	3	-4	-9	0

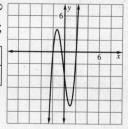

Review and Assessment *continued*

7. $x^3 - 2x^2 + 2$ **8.** $x^2 - 7xy + 12y^2$

9. $2x^3 + x^2 + 1$ **10.** $(10x - 3y)(10x + 3y)$

11. $(y - 1)(y^2 + y + 1)$

12. $5xy(3x^2y^2 + 2xy + 1)$ **13.** $9, -9$

14. $-6, 0, 1$ **15.** $0, 20$ **16.** $x^2 - 4x - 12$

17. $2x^2 + 5x + 3$

18. Possible zeros: $1, -1, 5, -5$; Zeros: $1, 5$

19. Possible zeros: $1, -1, 2, -2, 4, -4, 8, -8$

Zeros: $-4, 1, 2$ **20.** $x^2 + x - 20$

21. $x^3 - 3x^2 - 4x + 12$ **22.** $4.66, 2.80, -1.75$

23. x-intercepts: $(-2, 0), (2, 0)$; local max: $(0, 4)$

local mins: $(-2, 0), (2, 0)$

The graph rises to the right and to the left.

24.

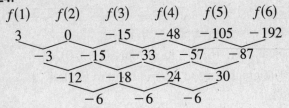

$$
\begin{array}{cccccc}
f(1) & f(2) & f(3) & f(4) & f(5) & f(6) \\
3 & 0 & -15 & -48 & -105 & -192 \\
& -3 & -15 & -33 & -57 & -87 \\
& & -12 & -18 & -24 & -30 \\
& & & -6 & -6 & -6
\end{array}
$$

Test C

1. $\dfrac{1}{x^3y^2}$ **2.** x^5y^5 **3.** x^3y^3 **4.** x^4y^{14}

5. $f(x) \to -\infty$ as $x \to -\infty$

and $f(x) \to +\infty$ as $x \to +\infty$;

x	y
-2	0
-1	6
0	4
1	0
2	0

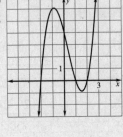

6. $f(x) \to +\infty$ as $x \to -\infty$

and $f(x) \to +\infty$ as $x \to +\infty$;

x	y
-2	4
-1	0
0	6
1	4
2	0

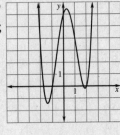

7. $-x^3 + 6x^2 - 2x + 6$ **8.** $x^2y^2 + xy - 12$

9. $2x^3 + 3x^2y + 3xy^2 + y^3$

10. $4(2x + y)(2x - y)$

11. $(2y + 1)(4y^2 - 2y + 1)$

12. $4(c + d)(c - d)(c + 2d)$ **13.** $6, -6$

14. $0, -3, 4$ **15.** $\frac{1}{2}, \frac{3}{2}$ **16.** $x^2 + x + 3$

17. $x^3 + 3x^2 - x - 1$

18. Possible zeros: $1, -1, \dfrac{1}{2}, -\dfrac{1}{2}$; Zero: $-\dfrac{1}{2}$

19. Possible zeros: $1, -1, 2, -2, 3, -3, 4, -4,$

$6, -6, 12, -12$; zeros: $-4, -1, 3$

20. $x^3 + 6x^2 + 11x + 6$ **21.** $x^4 - 5x^2 - 36$

22. $9.67, 1.93, -1.61$

23. x–intercepts: $(-4, 0), (4, 0)$;

local max: $(0, 16)$; local min: $(-4, 0), (4, 0)$

The graph rises to the right and to the left.

24.

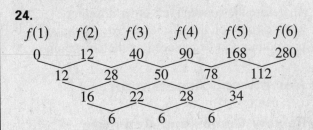

$$
\begin{array}{cccccc}
f(1) & f(2) & f(3) & f(4) & f(5) & f(6) \\
0 & 12 & 40 & 90 & 168 & 280 \\
& 12 & 28 & 50 & 78 & 112 \\
& & 16 & 22 & 28 & 34 \\
& & & 6 & 6 & 6
\end{array}
$$

SAT/ACT Chapter Test

1. C **2.** A **3.** D **4.** D **5.** B **6.** C **7.** B

8. A **9.** C **10.** A

Alternative Assessment

1. a–b. Complete answers should address these points: **a.** • Explain that complex conjugates are zeros that come in pairs $a + bi$ and $a - bi$, where a and b are real numbers. The polynomial function $f(x) = x^4 + 5x^2 - 36$ has two zeros that are complex conjugates, $3i$, and $-3i$.

b. • Explain that the polynomial must be of degree 3. If $i\sqrt{7}$ is a zero, then $-i\sqrt{7}$ must also be a zero. With three zeros, by the Fundamental Theorem of Algebra, the polynomial would be of degree 3.

2. a, b. $4x^3 - 8x^2 - x + 2$ **c.** • Answers will vary. **d.** • *Sample answer*: Synthetic division is used when dividing a polynomial by an expression of the form $x - k$. An example where synthetic division would not be used is $(4x^4 - 2x^2 + 3x - 1) \div (2x^3 + 4x - 1)$.

Review and Assessment *continued*

e. • The remainder is zero. This means that $(x - 3)$ is a factor of the expression $4x^4 - 20x^3 + 23x^2 + 5x - 6$.

f. • $\pm 6, \pm 3, \pm 2, \pm 1, \pm\frac{3}{2}, \pm\frac{3}{4}, \pm\frac{1}{2}, \pm\frac{1}{4}$

3. *Sample answer:*

$2x^3 + 3x^2 + 4x + 5 + \dfrac{16}{2x - 3}$; It is possible to get this result using synthetic division. Factor a two out of the quotient and the divisor to get a leading coefficient of one for the divisor. Thus, the problem becomes

$$\left[2\left(2x^4 - \frac{1}{2}x^2 - x + \frac{1}{2}\right)\right] \div \left[2\left(x - \frac{3}{2}\right)\right].$$

After cancelling the twos, proceed with synthetic division as normal.

Project: Playing the Game

1. *Sample answer:* Questions related to polynomial functions are written on index cards. Each question is rated 1 to 6, with 6 being the most difficult. Players take turns rolling a number cube and choosing a question with the appropriate rating. If the player answers the question correctly, the player receives the number of points shown on the number cube. The first player to receive 40 points wins. **2.** Check students' work. Questions can be either specific problems related to a given polynomial (e.g. solve an equation or factor a polynomial), or they may be general, conceptual questions. Students could create a game where polynomials are used to generate numbers in the game, such as evaluating a polynomial to determine how many moves to take in a board game. **3.** Check work. **4.** Check work.

Cumulative Review

1. -14 **2.** -5 **3.** 14 **4.** -26 **5.** -27
6. -14 **7.** 6 **8.** $-\frac{19}{2}$ **9.** $\frac{23}{11}$ **10.** 6 **11.** 8
12. 10

13. $x \geq 2$

14. $x > -2$

15. $x > \frac{3}{2}$ or $x < -3$

16. $x > 5$ or $x \leq -5$

17. $-2 < x < 3$ **18.** $-2.5 \leq x \leq -0.5$

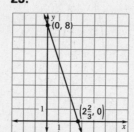

19. Line 2 **20.** Line 2 **21.** Line 1 **22.** Line 2
23.

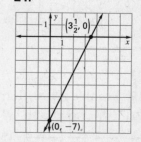

24.

25.

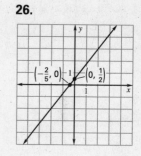

26.

27.

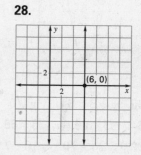

28.

29. $y = 3x + 5$ **30.** $y = \frac{2}{5}x - \frac{24}{5}$
31. $y = -\frac{3}{4}x + 5$ **32.** $x = 2$
33. $y = -\frac{3}{2}x + 2$ **34.** $y = -\frac{2}{3}x + \frac{7}{3}$ **35.** 2
36. 6 **37.** -3 **38.** 0
39. infinitely many solutions **40.** none
41. one **42.** $(3, -2)$ **43.** $\left(\frac{1}{2}, 4\right)$ **44.** $(1, 0)$
45. $\left(\frac{1}{3}, 4\right)$ **46.** $(-2, 5)$ **47.** $(0.5, 0.3)$ **48.** -8
49. -6 **50.** 7 **51.** -65 **52.** -14 **53.** 1

Review and Assessment *continued*

54.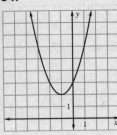

55.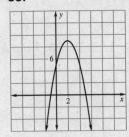

56.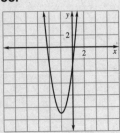

57.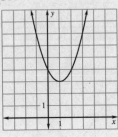

58.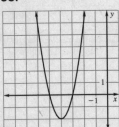

59.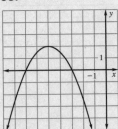

60. $\pm\sqrt{3}$ **61.** $3\pm\sqrt{6}$ **62.** ±4 **63.** $-\frac{3}{2}$

64. ±3 **65.** $-\frac{2}{3}, 5$ **66.** 5 **67.** $2\sqrt{5}$ **68.** $\sqrt{2}$

69. $\sqrt{26}$ **70.** $\sqrt{10}$ **71.** $\sqrt{11}$

72. $\dfrac{-1\pm\sqrt{41}}{4}$ **73.** $\dfrac{9\pm\sqrt{41}}{20}$ **74.** $4\pm\sqrt{7}$

75. $\frac{4}{3}, -1$ **76.** $\pm3\sqrt{2}$ **77.** $-2\pm\sqrt{3}$

78. $(3x^2 + 1)(3x^2 - 1)$ **79.** $(x^2 + 2)(x^2 + 1)$

80. $3x(2x^2 + 1)(x^2 + 2)$

81. $3x(x - 3)(x^2 + 3x + 9)$

82. $(4x^2 + 1)(x^2 + 9)$ **83.** $(3x^2 + 1)(2x + 3)$

84. $2x^2 + 4x + 5 + \dfrac{15}{x - 2}$

85. $x - 1 + \dfrac{-4}{x - 1}$ **86.** $3x + 8 + \dfrac{38}{x - 4}$

87. $x^3 + x^2 + 3x + 10 + \dfrac{5}{x - 3}$

88. 8 in. on each side